THE WANTAGE TRAMWAY

THE WANTAGE TRAMWAY

THE STORY OF A BUCOLIC ROAD SIDE STEAM TRAMWAY

RICHARD MARKS

PEN & SWORD TRANSPORT

AN IMPRINT OF PEN & SWORD BOOKS LTD.
YORKSHIRE – PHILADELPHIA

First published in Great Britain in 2024 by
Pen and Sword Transport
An imprint of
Pen & Sword Books Ltd.
Yorkshire - Philadelphia

ISBN 978 1 39904 834 7

A CIP catalogue record for this book is available from the British Library.

Typeset in 11/14 Palatino by SJmagic DESIGN SERVICES, India.

Printed and bound in the UK on paper from a sustainable source
by CPI Group (UK) Ltd., Croydon. CR0 4YY.

Pen & Sword Books Ltd. incorporates the imprints of Pen & Sword Books: After the Battle, Archaeology, Atlas, Aviation, Battleground, Discovery, Family History, History, Maritime, Military, Politics, Select, Transport, True Crime, Fiction, Frontline Books, Leo Cooper, Praetorian Press, Seaforth Publishing, Wharncliffe and White Owl.

For a complete list of Pen & Sword titles please contact

PEN & SWORD BOOKS LIMITED
George House, Beevor Street, Off Pontefract Road, Hoyle Mill, Barnsley,
South Yorkshire, England, S71 1HN.
E-mail: enquiries@pen-and-sword.co.uk
Website: www.pen-and-sword.co.uk

or

PEN AND SWORD BOOKS
1950 Lawrence Rd, Havertown, PA 19083, USA
E-mail: uspen-and-sword@casematepublishers.com
Website: www.penandswordbooks.com

MIX
Paper | Supporting responsible forestry
FSC® C013604

CONTENTS

ACKNOWLEDGEMENTS

I would like to thank all of the staff at the Royal Berkshire Archives for all of their help, so readily given, in finding all of the surviving Wantage Tramway Company records. The majority of the company archives which survived the closure of the company in 1946 were passed to the Berkshire County Archives since the town was located in the county before the boundary changes in 1974. The mass of surviving documents has been consulted in its entirety, providing a huge amount of new primary evidence to support the creation of this history of the tramway.

I would like to thank all of the staff at The National Archives in Kew for their assistance with the tramway archives held there. I would also like to thank Trevor Hancock and Suzie Tilbury at the Vale and Downland Museum in Wantage for their help and advice on the history of Wantage which added context to this book and allowed it to become more than it otherwise could have been. The help received from Katie Amos at Reading Central Library and Peter Rance at the Great Western Society in Didcot has been invaluable and I am grateful for their assistance in putting the research together for this book.

I would also like to thank John Scott-Morgan not only for commissioning this book, but also for his advice and assistance during the research and production of this volume as well as allowing me access to his image collection for use in this volume.

Finally, I would link to thank Ros and Lexi for their help, patience and support during the research and creation of this book.

Cover images from the Vale and Downland Museum and John Scott-Morgan collections

INTRODUCTION

The construction of Britain's railway network was virtually complete by the late 1870s, with most of the major trunk routes built and operating. The only main trunk route that remained to be added to the railway map was the Great Central's line into Marylebone. The railway manias of the mid-nineteenth century did not see every place in Britain connected to the railway. Many market towns which were once important saw the railways bypass them due to the need to avoid expensive diversions and civil engineering works.

Railway engineers surveyed their routes to achieve the most effective line serving the main towns on the route, but also with a mind to fast running between the major nodal points on the railway. The Great Western's main line between Bristol and London was constructed with hub stations at the major towns such as Reading and Bath. The intermediate market towns were bypassed in the main, unless they were located on the geographically favourable route surveyed by Brunel. These rural places were not completely abandoned, though.

Many of the Victorian railway companies adopted the use of 'railheads' in rural areas, the Great Western Railway particularly notable for its 'Road' stations which served exactly that purpose. Located in places which appeared to serve no one, these stations were at most three miles from the majority of local settlements and many large farms. Whilst not entirely convenient, the local 'Road' station provided a good enough railway link to the local population and businesses, while allowing the railway company to construct a viable line.

In the nineteenth century, the concept of walking for an hour to get to a different place was quite common and would allow someone to reach their local railway. The three miles or so between railhead and town or farm could be covered more quickly by horse and cart, farmers being able to deliver produce to their nearest station whilst also collecting any inwards goods such as machinery, seeds and fertiliser. The facilities which were initially provided by the railways might have been acceptable at first, but as the nineteenth century progressed the population of many towns began to look at building their own railway to connect with the major rail arteries. These schemes were often devised in the hope that the local railway company would purchase or at least operate the branch railway and would provide more convenient services for local passenger and goods traffic.

One such town was Wantage. The Great Western Railway's (GWR) Paddington to Bristol line bypassed the town, passing three miles away. The GWR opened their Wantage Road station in 1846, serving the local area as a typical Victorian rural railhead. Wantage and most of the local villages and farms lay at most around four miles or thereabouts from the station. Whilst initially adequate for the people and businesses of Wantage, the townsfolk, wanting a better and more convenient rail connection, began to make plans for a new company which would meet their transport needs through a link with the new railway.

Companies, and particularly railway companies, were often formed in the nineteenth century by a small number of promoters who would hope to drive the project through to completion. The companies were funded by shareholders, many of whom were local people in the case of short branch lines. Whilst many schemes failed to gain enough support and were unable to achieve the critical mass to gain the Act of Parliament which was required in order to construct a railway, some were successful. For Wantage, this was one of the cases where the expensive railway line did not gain support.

The promoters of a railway would attempt to gather local support for the scheme through meetings in the places which the planned railway was hoping to serve. Often the nascent company would enlist a notable local person, although these people were frequently those promoting the railway, to provide potential investors with more confidence in the scheme. In order to raise the capital needed to cover the considerable expense of building even a short branch line railway, the promoters would need to first get agreement that a company should be formed and, once created, for the company to be capitalised through the issues of shares in the venture.

The amount of shares and value of each share had to be agreed at the meeting when the company was formed, and at the same time a chairman and a board of directors had to be appointed from within the group attending. Once done, the share offer could proceed, but no monies would change hands. The shares were distributed through a subscription agreement whereby an individual would commit to provide capital up to the value of a subscribed number of shares. The subscribers would receive a certificate, not for shares since they did not own those yet, but for the total that they had agreed to fund. This certificate, a scrip, could not officially be traded, but often were.

The directors could call upon the subscribers to pay varying amounts of their commitment up to the full value of the shares they had agreed to purchase. Once the value of the shares had been paid to the company in full, the subscriber became a shareholder of the company and was then eligible to receive dividends from the company's profits, as decided by the directors, and to vote at annual and extraordinary general meetings. The shares could then also be sold and if a person owned an agreed quantity of fully paid-up shares, they would be able to become a director of the company. The Wantage Tramway was no different, the directors being elected from the major shareholders.

The Victorian railway companies were funded through two different types of shares, both of which were important to the Wantage Tramway Company. The basic share was made available to fund the expenses associated with the building of the line, and, if necessary, the purchase of rolling stock, horses and locomotives, as well as other equipment required to run the line. No guarantees were given of return on the investment, the company deciding upon a variable dividend to be paid each year as a result of trading profits. A proportion of trading profits, often agreed at the annual general meeting of the company attended by shareholders, was used to pay a dividend as a percentage of the value of the shares held, perhaps 5 per cent in a good year. In bad years, the company would potentially not pay any dividend at all.

The second type of share, which the tramway resorted to in order to resolve short term capital shortages or to make investments in new equipment, was the preference share. These shares were used to raise capital quickly and came with a guaranteed return. The holder of preference shares was entitled to a dividend from profits each year at the percentage rate at which the shares had been offered. These were taken from the profits before any dividends were paid on ordinary shares, which occasionally meant that only preference shareholders received a return. There was, however, a risk associated with these shares. In good years, a company may be able to pay a higher percentage dividend on ordinary shares than the guaranteed rate that was paid on preference shares.

Railways required a lot of capital to build and open, but a change to legislation would come to the aid of the people of Wantage. In 1870, the British government passed the Tramways Act, which changed the requirements which had to be met to build a railway. A railway project involved the expense of going through the railway committee and all the other costs associated with applying and gaining an Act of Parliament authorising the building and operation of a railway. The level of civil engineering that was required in order to meet the requirements of the Railway Department of the Board of Trade (BoT) was both costly and time consuming to deliver, but the Tramways Act offered an alternative.

Instead of applying for a full railway act through parliament, the proposers of a tramway could now apply directly to the BoT in the first instance for a 'Provisional Order' which allowed them to begin construction and then create the company to run the line. The plan and section of the route with the book of reference, which covered the property that was to be purchased to make the railway, still had to be deposited with Clerks of the Peace of the counties through which the line was to run. The Tramways Act allowed the company to construct their line along the middle of roads, but this may have precluded the use of mechanical propulsion on the tramway until the act was later amended to permit its use. The default power covered by the Act was to be animal power alone, but this would change and would be driven partly by the Wantage Tramway.

The promoters of a tramway were required to advertise their intention to apply for a Provisional Order in the October or November before the application was to be submitted. The advertisements were to be placed in either the local or county papers related to the areas in which the line would be built on two successive weeks, and by poster along the intended route. The notification was also to be placed for two weeks in either the *London* or *Edinburgh Gazette*.

The BoT required that the advertisements should contain the objective of the planned application, a general description of the works that would be required to build the line and a list of where these were to be undertaken. The proposers had to also include in the advertisements when and where the plans would be deposited and could be inspected by the public, and finally an office address where correspondence could be sent.

Proof that the advertisements had been placed had to be supplied to the BoT no later than the 15th day of the month following their placement. The company then needed to serve notice to both the House of Commons and House of Lords that they were planning on applying for the order before 15th December of the same year.

The new act had one similarity with the railway acts, in that the BoT would determine the byelaws of the tramway and the traffic which the line was to convey, passenger, general freight, heavy goods, parcels, or any combination thereof. The board would also set the maximum tolls and charges which the tramway was able to levy, any increase above that requiring an application to the department for permission to increase the charges being made. A security was required of no less than 4 per cent of the cost of the entire project, which the promoting company was to deposit with the Court of Chancery. The Court would, however, invest the money on behalf of the company, and upon successful completion of the scheme would refund the cash with the interest that it had earned whilst on deposit with them. Once the deposit had been made, the board would provide the company with their Provisional Order which allowed them to construct, but not open, the tramway lines.

The services of a parliamentary agent would still be needed to ensure that the documentation required was placed with the relevant bodies in Westminster, and to act as the tramway's representative to the BoT, and both Houses. The workload and costs associated with the process would, however, be far less than for a full railway line.

Once all the provisions had been met, the BoT would, after at least seven days following the submission of proofs but no later than 25 April following the application, arrange for a Bill to be introduced to the Houses of Commons and Lords. The Bill, when passed, would convert the provisional order to a full Act of

Parliament authorising the tramway. Typically, a number of tramways were covered by one Act, which saved cost and time. The Act did not allow the tramway to operate until such time as the line and the rolling stock to be used upon it had been passed as fit by a BoT inspector. A successful inspection would result in the BoT granting a licence for the tramway to operate and the final byelaws under which that licence would apply.

The BoT Rules relating to the granting of the tramway order required that the promoters of the tramway should create a limited company incorporated under the Companies Act, 1862. The tramway was restricted in its construction with a space of no less than 9ft 6in being provided between the nearest rail and the footpath on either side of the line, but if it was intended to run vehicles 'adapted to run on railways' a space of 10ft 6in was required. Lastly, the BoT rules required that the tramway company should give the board at least fourteen days' notice of their intention to open the line and an inspection by a nominated BoT inspector must have been undertaken and passed.

The Tramway Act offered a cost-effective means by which the promoters of a railway could build their planned route, especially in areas where traffic would be relatively light or where a short, lightly built line would provide an advantageous link to a local mainline railway. The Act met the needs of the people of Wantage perfectly, a tramway being an ideal solution to the need to connect the town more conveniently to the GWR at Wantage Road station.

The tramway was not the first attempt to connect Wantage to the GWR, though. A proposal to build a full railway branch line from Wantage Road was promoted some time before the Tramways Act was passed. The proposal would not come to fruition.

Chapter 1

THE GREAT WESTERN RAILWAY AT WANTAGE ROAD

The GWR was not the first mass transport system to reach the neighbourhood of Wantage. The Wilts and Berks Canal opened a branch to the town which terminated in Mill Street directly opposite what would become Clark's Mill. The canal branch served the town well, bringing in supplies of coal, barley and iron ore which supported the town's growing industries.

The GWR's trunk line from Paddington to Bristol was surveyed to take the route needing the least heavy earthworks which supported fast running. The railway company did not think it necessary to directly serve every town that their route was close to, as the diversions into each place would add cost to the project which was viewed as unnecessary. The company adopted a programme of rail heading for places where the potential traffic was viewed as not likely to justify bringing the line into a town. The final route surveyed for the GWR line west of Reading was designed to provide the fastest and most direct link possible, with minimal expensive civil engineering works, between Bristol and London. The resulting 40 miles of fast running after Reading, whilst good for the GWR, meant that wayside towns missed out on a direct railway connection.

In Berkshire, a lot of previously important market towns were bypassed by the railway as it made its way through the Berkshire Downs. Faringdon, Wallingford and Wantage were all left remote from the early Victorian railway network, as was Abingdon. The reasons for the latter's disconnection was entirely different and more caused by local political influence than railway economics.

The three towns were served by the GWR through one of their remote 'road' stations. Wallingford was initially served by Moulsford station which opened in 1840, and in October was renamed Wallingford Road. The station was located around three and a half miles from the town it purported to serve. Faringdon received a similar service through another of the GWR's stations in a remote location. Faringdon Road was located near Challow, more conveniently for Wantage, which was 4 miles distant, than for Faringdon, which was a 6-mile journey from the rail hub which claimed to serve it. The three towns at least had a rail link within a distance acceptable to Victorian expectations.

The GWR's decision to choose to site its 'Faringdon' station between both towns appears logical, given the railway's use of hubs in rural locations, with one facility serving both through the well maintained turnpike road from Challow to Wantage and Faringdon. The later Uffington Junction station, four miles from Faringdon, opened when the Faringdon Railway built its branch from the GWR in 1864. The GWR renamed Faringdon Road to Challow in 1864 following the opening of the

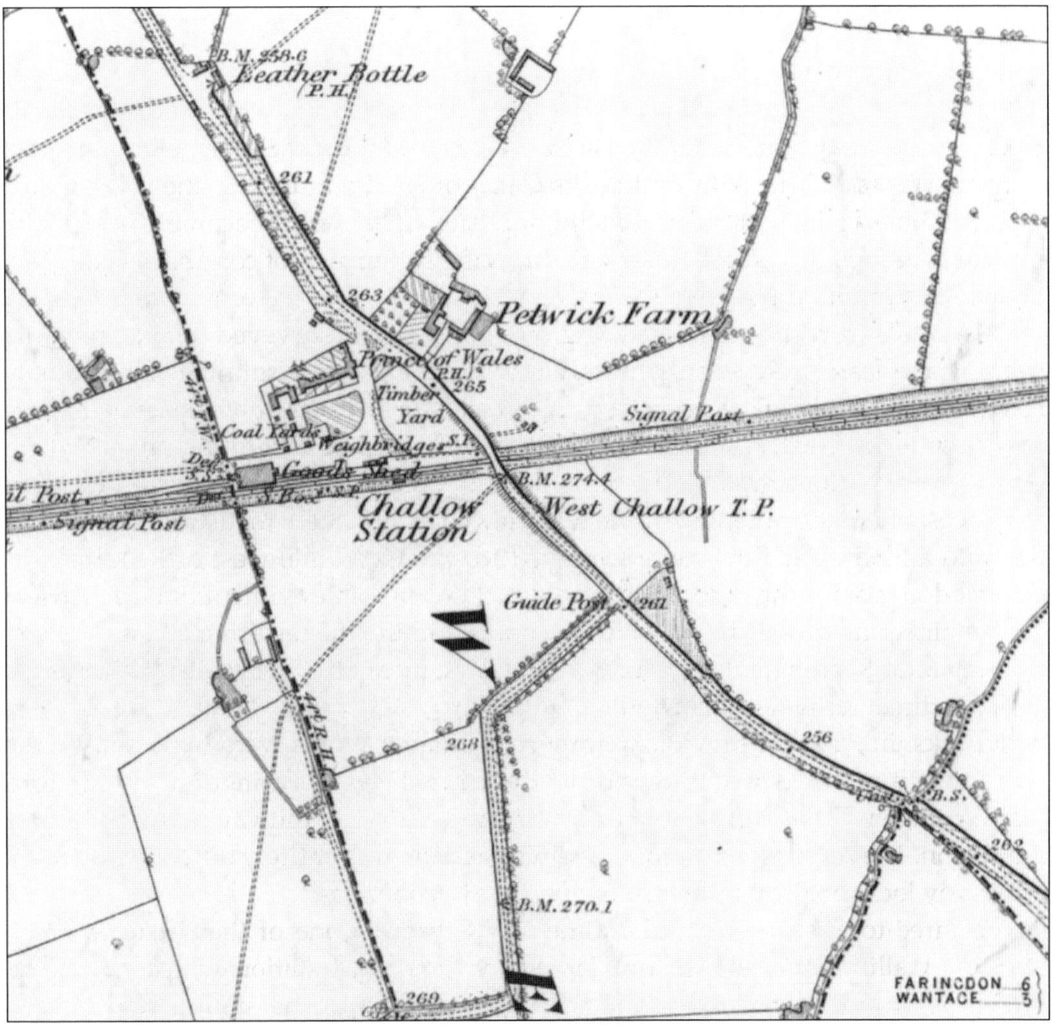

Challow station, formerly Faringdon Road. Faringdon lies to the north east along the main road whilst Wantage is to the south east along the same road. The location of the station in a rural area is a notable feature of the GWR's 'Road' stations. *Excerpt of Ordnance Survey Six-Inch Series, Berkshire Sheet XII, 1883.Reproduced with the permission of the National Library of Scotland*

branch line, to avoid confusion. The new name finally reflected the location that the station served.

Wantage remained reliant upon Faringdon Road station until 1846, when the Great Western opened its new station, Wantage Road, on the Besselsleigh Turnpike Road (now the A338). The opening of the new station gave the town two rail hubs to choose from, although it seems that Wantage Road station, a mile closer to the town than Faringdon Road, was preferred by most of the population of the town.

The station at Wantage Road was a modest affair served by a small number of stopping trains each day. It consisted of separate, single faced, Up and Down platforms serving the two track broad gauge main line. The lines were crossed by a single arched brick bridge carrying the main road over the railway which was raised at this point in order to allow trains to pass under it. A three-storey station building was provided at the London end of the Up platform immediately adjacent to the main road. The building opened from the ground floor onto the platform, however the entrance to the station was on the first floor level, serving the raised main road on the extended bridge abutments. The ticket office appears to have been at road level rather than at platform level, with the building nestling

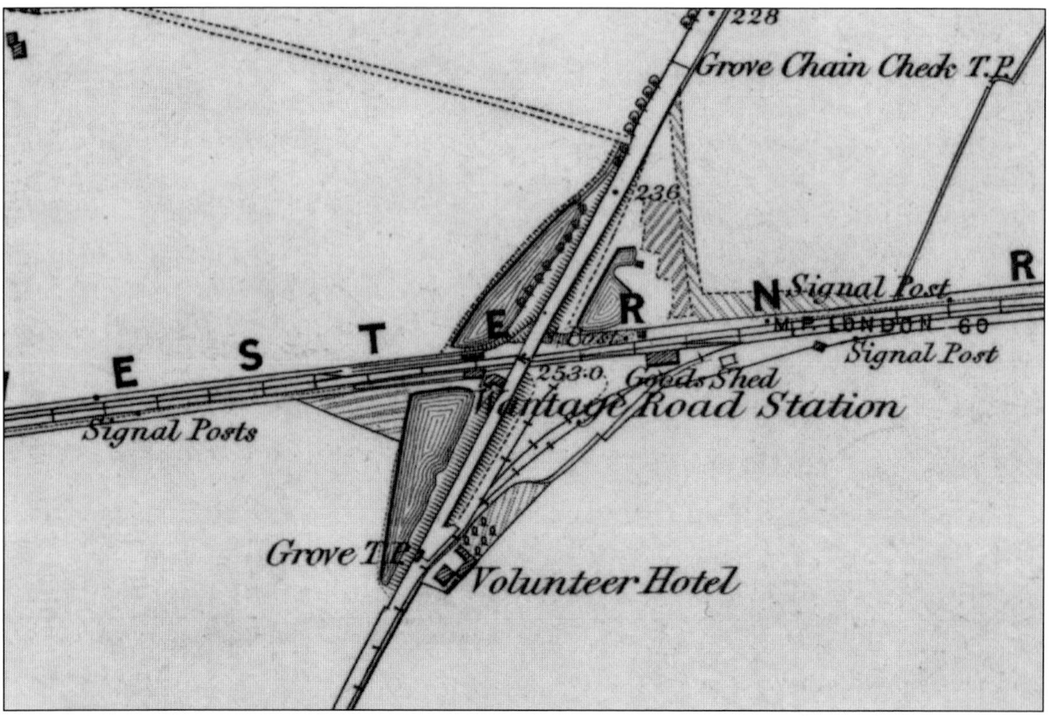

Wantage Road Station in 1883, showing the tramway sidings which had been added since the opening of the station. *Excerpt of Ordnance Survey Six-Inch Series, Berkshire Sheet XIV, 1883. Reproduced with the permission of the National Library of Scotland*

into the embankment built to raise the road above the railway. The 1883 Ordnance Survey map shows no crossing between the Up and Down platforms, but it can be assumed that a barrow crossing was provided since there seems to have been no access from the road onto the Up platform. The Great Western provided a footbridge at the London end of the station at some point after the opening of the station, which shows in the 1900 Ordnance Survey.

Each platform was provided with a simple wooden passenger shelter with a small platform canopy extending from its front. The shelter on the Up line platform, serving trains towards Reading and London, shows on a 1910 image as fully enclosed whilst that on the Down platform for Swindon and Bristol-bound trains was open on the side facing the rails. Few other facilities seem to have been provided for the passenger on the platforms at that time.

The GWR provided a single siding next to the Up line immediately west of the Down platform into which goods trains could be shunted if required to make way for faster trains. The GWR's goods facilities at the station consisted of a single siding running parallel to the Down line starting immediately next to the London side of

Hughes tram, No. 4, in Wantage Road yard with passenger car No. 3. The bar coupling can be seen between the engine and the tramcar. The building just visible at the top left is the GWR's station building with the first and second floors visible, the ground floor is behind the embankment leading to the bridge behind the tram car. The narrow archway directly behind car No. 3 led onto the down platform of the GWR's station. *John Scott-Morgan Collection*

the road bridge with a goods shed situated next to it. The siding continued some way east of the shed to provide additional wagon space. The siding appears to have only been accessible through a pair of trailing points from the Down line located near each end of the goods shed. The ordnance survey maps show no access to the shed from the Up line and with no handling facilities provided for the Up line siding it seems unlikely that goods were dealt with there.

The Great Western provided an access arch within the road bridge adjacent to the Down running line separated from the main arch by a brick pier. The archway which initially allowed railway staff to walk between goods yard and station directly, without having to walk along the railway itself, would later provide an access path for Wantage Tramway passengers to move freely and safely between the tramway station and the Great Western's platforms.

In the early days of the railway service, there was no direct rail link into Wantage town itself. In order to link the town to the railway, a means to move freight between the town and the station would be needed. The question of how to transport the passenger arriving at Wantage Road also needed to be addressed. The provision of onward transport for people and goods would be beneficial to the railway and given, that the Victorians were keen to provide an integrated transport system where possible, a solution was desirable to the company directors and management. As a rural wayside station, Wantage Road did not have a high level of staffing, and unlike larger stations in the main Berkshire towns was not provided with delivery carts. The railway had another means to resolve the problem which they commonly used for their smaller stations.

The GWR entered into contracts with local businesses in order to provide onward road transport from their rural stations where the company did not have its own road service. The company had a standard preprinted contract agreement which simply needed to be completed by the goods manager's office at Paddington and authorised by the company lawyers in the GWR's legal office in Westminster. The agreement, once signed by the subcontractor, would form the contract between railway and agent.

These contracts covered both delivery to the consignee of the parcels and goods as well as allowing the companies who held them to collect goods on behalf of the railway company for delivery to the station for onward dispatch by rail to their final destinations. The onward transport of passengers from the GWR's rural stations was often covered by a separate agreement. The contracts gave the companies concerned the sole, or joint rights in the case of multiple agreements, to transport passengers or goods from or to the station. One interesting point that can be found in the agreement documents was a clause that the company concerned must clearly display on all vehicles used for onward transport that the company were agents for the Great Western Railway Company. The terms of the standard GWR agreement would later impact the tramway when it took over the running of onward services into Wantage town.

Once the railway station was open at Wantage, agency agreements were drawn up in the usual manner by the GWR at Wantage Road. Two local companies were given agency contracts which allowed them the sole rights for the onward transport of goods and passengers from Wantage Road into the town. These agencies still remained in force at the time the tramway company was formed. Passengers and light parcels were carried into Wantage using the omnibus belonging to Thomas Nunney, who was also the proprietor of the Bear Hotel in Market Place, Wantage. Goods traffic, excluding parcels, was the sole responsibility of Samuel Sims of Grove Street in Wantage, who held the agency from the Great Western to cover the onward delivery of freight from Wantage Road. The agreement covered the collection of freight from Wantage and the surrounding area, for delivery into the care of the GWR at the goods yard at Wantage Road. Samuel Sims' son James later took on the operation of the agreement with the GWR.

The agency agreements did not restrict private individuals or businesses from making their own arrangements to reach the railway station. In Wantage, the Vale of The White Horse Foundry used their own traction engine to take their manufactured products directly from the foundry to the railway station and continued to do so into the twentieth century. The decision would cause the company some problems in the future, but for most of the nineteenth century, the traction engine served them well. The railway benefited later in the century from a large amount of traffic generated by another local engineering company, Nalder and Nalder, who had a considerable role to play in the Wantage Tramway after it had opened. The company was located in Challow and so was more likely to have used Faringdon Road station as it was probably more convenient for them.

The town appeared to be adequately served by the two stations at Wantage Road and Faringdon Road for some time. The next twenty years passed with the townspeople using the GWR's services for travel, the import of goods and raw materials into the town, and exports out of finished products and agricultural produce. The canal remained an important artery for less time-sensitive shipments and it seems that, at least for a time, the combination of the existing railway and canal, as well as the local turnpike roads, served the town's needs.

The situation began to change in the 1860s as business was growing and the people of Wantage began to demand a more convenient transport offering. Thoughts began to turn to bringing the railway into the town itself, with consideration of the possibility of the creation of a branch line from Wantage Road. The GWR was unlikely to entertain the construction of the short line themselves since the town had already been bypassed and so thoughts turned to the creation of a local independent railway company to potentially undertake the construction of a railway.

Chapter 2

THE WANTAGE AND GREAT WESTERN JUNCTION RAILWAY

The Wantage Tramway was not the first project which aimed to bring the railway directly into Wantage town. A group of promoters looked to create a branch from Wantage Road station on the GWR's mainline to Bristol, proposing a branch railway from Wantage Road during the 1860s. The scheme was not promoted by local people but by a group of men who seemed to have an interest in building rural branch lines. The proposers of the railway consisted of George Reed of Burnham, Buckinghamshire; Owen Strong of Iffley near Oxford; John Fredricks of Abermellte in Glamorgan; Edward Roberts of Westminster, London; and George Newington of Staplehurst in Kent. George Reed was also a director and shareholder of the Abingdon Railway Company.

The proposals for the railway appear to have proceeded favourably, the work undertaken by the proposers seemingly indicating that many shares had been subscribed. A company was thus formed to construct the railway and was granted an Act of Parliament on 16 July 1866 which authorised the railway under the auspices of the Wantage and Great Western Junction Railway Company. The railway was authorised to build their branch which was to be 2 miles, 2 furlongs and 5 chains in length, beginning at the rear of the Down passenger platform, which was to become double sided with the new side serving the branch line as could be seen at Maidenhead with the opening of the Wycombe Railway or at Cholsey when the Wallingford branch opened. A terminus was to be provided for the branch in Mill Street in what was then the garden of a house owned by William Southby.

The branch line was to form a junction with the GWR main line, joining the Down line 6ch from the end of the platform. An additional clause, which would be seen again when the tramway was formed, was added to the Act. The clause gave the GWR complete control over how the junction would be constructed. It was to be built according to an agreement to be made between the branch company and the GWR, whereby the line could only be opened once the Great Western's engineer was satisfied that the junction had been built to the standard they demanded.

The railway, as was always the case in Railways Acts, had to be built within the plans submitted to accompany the application for the Act which authorised it. The plans contained a line of boundaries either side of the intended line, referred as the 'limits of deviation'. The railways were authorised to build within these limits to allow the line to be slewed to avoid springs, unfirm ground, or just to provide smoother curves onto bridges and into stations. If the railway were to find that there was a problem with where the line had been surveyed and which had not been foreseen by the engineer, the company would have to go to the expense of applying for a new Act of Parliament to change the authorised route.

The plans and sections of the proposed railway, as well as the Book of Reference which nominated the properties which were to be purchased in order for the railway to be built, had been completed and deposited with the Clerk of the Peace for Berkshire and in Westminster as required for the granting of a Railway Act. The completion of the survey and plans and the appointment of an engineer to oversee the construction of the line points towards a serious plan to build and open the railway.

The plans submitted show the line leaving the GWR at the rear of the Down platform and proceeding parallel to the Besselsleigh Road across farm land before crossing the Wantage to Grove road through a level crossing. The line was then to have continued across the Wilts and Berks Canal by a bridge before heading towards Wantage across farm land. As the line was about to enter Wantage, it would cross the River Ock three times on three separate bridges due to the bends in the river, before entering the town and terminating near Mill Street. A junction with the GWR was included on the plans which seems to deviate from that described in the Act. The Wantage and Great Western Junction was to have joined the GWR via a Y shaped junction, one arm of which took it into the branch platform and then onto the Great Western through the junction with the Down line. The other arm of the Y junction was to have turned away from the station and created a direct junction with the Great Western's Up line. It seems the promoters were anticipating the need for through goods and passenger traffic which avoided Wantage Road station.

The company was authorised to raise a total of £25,000 through the issue of 2,500 shares valued at £10 each, with the railway to be completed by 1869. The company was authorised to raise an additional £8,300 by mortgage if needed, but only after the entire share capital had been subscribed to, and more than half of the value of the shares had been paid up by the subscribers. Finance appears to have been sufficient to begin the project, as the Act of Parliament states that the promoters had deposited 8 per cent of the estimated cost of the scheme (£2,000) with the Court of Chancery. The deposit had been made in the names of Owen Strong and George Reed as subscribers of the undertaking. The promoters were clearly confident that the railway would be built, as the deposit would be forfeit if the railway was not built by the date specified in the Act unless the promoters could prove that they had called off and spent on works half the value of the share capital authorised by the act. The bond would also be forfeit if the railway were not built at all.

The Act of Parliament, as was usually the case, set down the maximum tolls and fares that the railway could charge for a variety of traffic. The charges were always split into two categories for each type of traffic. These were split according to whether the traffic was being carried on a vehicle owned by the passenger or consignor of goods, with an additional charge if the vehicle was owned by the railway. The Wantage and Great Western Junction was authorised to charge up to 6d per passenger, with an extra 2d if the carriage was owned by the railway. Passenger traffic was nearly always carried on a vehicle owned by the railway, although there were exceptions once in a while which the Act covered.

The Act did not make any allowance for charging by the mile, laying down only a flat charge for both passenger and goods traffic. In other Acts of Parliament, the tariffs are mostly laid out as a rate per mile allowing the company to create a scale of charges between stations on the line. The Wantage and Great Western Junction Act contained no such allowance, which shows that the company, at least at first, had no intention of creating an intermediate station on the line. The plan and section does not show any allowance having been made for any intermediate stops either. This may explain the comments later made by Edward Ormond during the meeting arranged in Wantage to create the tramway.

Goods traffic was segregated into 'classes' by type of goods. Class 1 traffic (horses, mules and other beasts of burden) was to be charged at no more than 8d per animal with an extra 2d for use of railway owned wagons. Class 2 (cattle) at 5d per animal (plus 2d for use of a railway wagon); Class 3 (small farm animals such as sheep or pigs) at 3d per animal (plus 1½d); Class 4 (dung, compost, undressed stone, coal) at 5d per ton (plus 1½d); Class 5 (iron, chalk, bricks, sand) at 6d per ton (plus 1½d); Class 6 (sugar, corn, nails, chains, light iron castings) at 7d per ton (plus 2d per ton); Class 7 (cotton, wool, drugs, manufactured goods, fish and all wares not covered by other classes) at 8d per ton (plus 2d).

Parcel tolls were restricted by weight of package below 56lb in weight. The company would have been able to levy a scale of charges if the railway had opened but were authorised to charge any toll that they thought suitable for any parcel over 56lb but under 500lb after which the goods tariff would supersede the parcel rates. The tramway would be restricted in the same manner when it opened, as rules for charging to carry traffic were also part of the tramways orders and authorising acts.

The new railway company was clearly being taken seriously in railway circles, as the Act stated that it would be 'expedient that the Great Western Railway Company should be empowered to work the said Railways' The GWR had obviously agreed with the promoters of the Wantage branch that they would be prepared to work the line on the company's behalf. It is likely that an arrangement similar to that agreed between the Great Western and the Abingdon Railway Company would have been put in place, since the GWR had a standard way of working with these lines until they absorbed them, which happened in the case of Abingdon Railway Company in 1904 and no doubt would have happened to the Wantage company had the railway been completed. The arrangements to operate lines where the GWR ran the trains for the owning company involved the Great Western taking an agreed percentage of all the revenue taken on the branch, the balance being paid to the owning company. It is likely this would have been the agreement for the branch at Wantage had it been opened.

At this point, the branch line appeared to have been progressing well and it seemed that the promoters had everything they needed. The GWR was involved in the project, an engineer was in place, capital was authorised and, it seems, subscribers had signed up to support the railway project. A mystery then emerges with the railway seeming never to have progressed beyond that point.

No construction seems to have been undertaken, but no Act of Parliament appears to have been passed authorising the abandonment of the railway, which was needed if the railway company decided, for whatever reason, not to proceed with construction of even a small element of an authorised scheme. The project to build a railway seems to have stalled, so a new scheme would now be required to link Wantage to the GWR by rail.

Chapter 3

PROPOSAL FOR A TRAMWAY

The failure of the Wantage and Great Western Junction Railway project left the town still without a direct rail link. The growth of the town's industries began to create a greater need for a more convenient transport link, as did the need for people to travel on business to local towns as companies began to expand their customer bases to a wider area. A railway was clearly needed in the minds of many in the town, but the prohibitive cost of building a branch line prevented a second attempt to build one. The passing of the Tramways Act in 1870 provided an alternative which was readily adopted by a group of local promoters who saw the opportunity to create a viable new transport link.

The promoters were in discussion with a group of potential investors after the passing of the Tramway Act in 1870, with a view to establishing a tramway company to construct and operate a link to Wantage Road station and to create a direct connection with the GWR. Thoughts of building a railway through Wantage had not entirely been abandoned though. Edward Ormond, by now nominated as the representative of the tramway proposers for correspondence, received a letter from J.F. Etherington Barnes, an experienced engineer based in Sherborne, Dorset. The letter, dated 3 January 1873, stated that Etherington Barnes was involved in a scheme which was shortly aiming to promote a railway from Wantage Road station to Wantage. Etherington Barnes asked Ormond for his opinion about whether a full railway scheme would be likely to gain support within the area. He also asked Ormond if, in his position as a solicitor, he had any clients along the proposed route of the railway who might be interested in supporting and investing in the railway scheme.

Etherington Barnes questioned Ormond about the nature of the land on the proposed route and also the likely level of traffic which might be expected by a railway company. The letter suggests that the new railway scheme's proposers had not surveyed a route themselves, although reference to a proposed route in the correspondence between Etherington Barnes and Ormond suggests that the proposers were intending to follow the route in the plan and sections created for the Wantage and Great Western Junction Railway.

The new railway, if built, would not have terminated at Wantage. The scheme was more ambitious than the Wantage and Great Western Junction had been. The new railway, according to Etherington Barnes, was to have continued past Wantage, extending as far as Lambourn before continuing to a junction with the proposed Swindon, Marlborough and Andover Railway at Marlborough. The proposal was to have a second leg, with a new railway to be constructed from Wantage Road station to Oxford. The railway, if built, might have provided a useful cross country cut off from Oxford, avoiding the need for trains to be routed through Didcot. It was not to be.

Correspondence between Ormond and Etherington Barnes continued, with Ormond clearly stating that a tramway was the preferred option for the town, although there was obviously a degree of bias in preferring the scheme in which Ormond was himself involved in proposing. A letter dated 7 January from Etherington Barnes clearly shows Ormond's position. Etherington Barnes suggested in the letter that the town would be better served by the railway scheme rather than a short tramway and that those interested in supporting the tramway should take up shares in the proposed railway company to the same value. Etherington Barnes stated that he believed that if the townsfolk of Wantage and the surrounding area were to back the railway, he could commit to build it. He also stated that he was sure that he would be able not only to convince the GWR to operate the line when it was finished, but also to invest in the project. Etherington Barnes finished by stating that 'his' railway would be the better project, and that it could be constructed throughout for a moderate cost per mile.

The correspondence ceases in the archive files after this letter, which suggests that Ormond refused to cooperate with the railway scheme and turned down the request to convince those interested in the tramway to instead invest in the railway scheme. The Wantage, Lambourne and Marlborough Railway project seems to have been abandoned at that point, joining a large number of other proposed railways which did not come to fruition in nineteenth-century Berkshire.

The group proposing the tramway quickly appointed an engineer to survey the route. George Stevenson of Wantage (not to be confused with the George Stephenson of Liverpool and Manchester Railway fame) was appointed and asked to conduct a survey of a proposed route which a new tramway might economically take between Wantage Road station and the town. Like many of the early engineers, Stevenson was not solely a consulting engineer, being listed in the 1854 Kelly's Commercial Directory for the town as an ironmonger based in Mill Street in Wantage. His entry in the 1876 Harrod and Co. Commercial Directory had changed to indicate he was now an 'ironmonger, gas engineer and contractor' mirroring how many 'engineers' were working.

Stevenson does not appear to have been a member of either the Institute of Mechanical Engineers (founded 1847) or the Institute of Civil Engineers (formed 1818). There was however, and still is, no statutory requirement for an engineer to be a member of any of the engineering institutions, although it would have been favourable in gaining work. It seems that Stevenson had not worked on any major engineering projects at the time the tramway was being proposed. His probable lack of formal training was not unusual for many of the early engineers, although formal university training had become available during the nineteenth century. His inexperience in laying down railways was partly the cause of some of the delays that occurred during construction of the line.

Nevertheless, Stevenson had surveyed his route, opting to closely follow mostly that which had already been surveyed and made publicly available for the defunct Wantage and Great Western Junction Railway scheme. At one point, Stevenson's route diverged from the previous one, crossing the northern end of his own garden, which he agreed to sell (for a nominal £5) to the tramway company. It is likely that the route surveyed for the railway scheme was the best route into Wantage and so it would be logical that Stevenson would have found the same line of rail would work equally as well for the tramway. The new route benefitted from the clauses in the Tramway Act which allowed it to run right alongside the road between Wantage Road station and the town, which the previous railway had needed to be separated from; nevertheless, the proposed tramway and railway routes were very similar.

With a route surveyed, the promoters called a public meeting for 'persons favourable to a Tramway Communication' to link the town with the railway. The meeting was scheduled for half-past three in the afternoon on Wednesday, 22 October 1873. The meeting was to be presided over by the then Colonel Robert Loyd Lindsay MP, giving the meeting the integrity to attract the public. Posters were placed around Wantage to publicise the meeting at Wantage Town Hall.

The meeting was well attended, showing the interest that the townsfolk had in improving their transport links. The chairman began by outlining what the proposers hoped to achieve through the construction of the tramway. The new line, if run well, was believed to be not only beneficial to the town, but also likely to be profitable and thus remunerative to the company's shareholders. Whilst it is likely that the promoters of every railway scheme, including the Wantage and Great Western Junction, would have made similar claims to justify investment in the project, it seems that on this occasion the claims were being made upon sound commercial foundations.

Edward Ormond took over the meeting after the chairman's initial address. His comments are interesting and seem to clarify why the previous junction railway scheme was not built. He seems to have suggested that, even though the Wantage and Great Western Junction had been granted its railway bill, the railway and the company formed to construct it was being run by 'strangers'. He suggested that the railway would be too expensive to build and was estimated to cost between £20,000 and £30,000. Ormond stated that he did not therefore advocate a railway, preferring a tramway, which he believed would be certain to pay an annual dividend to those who had invested in the scheme.

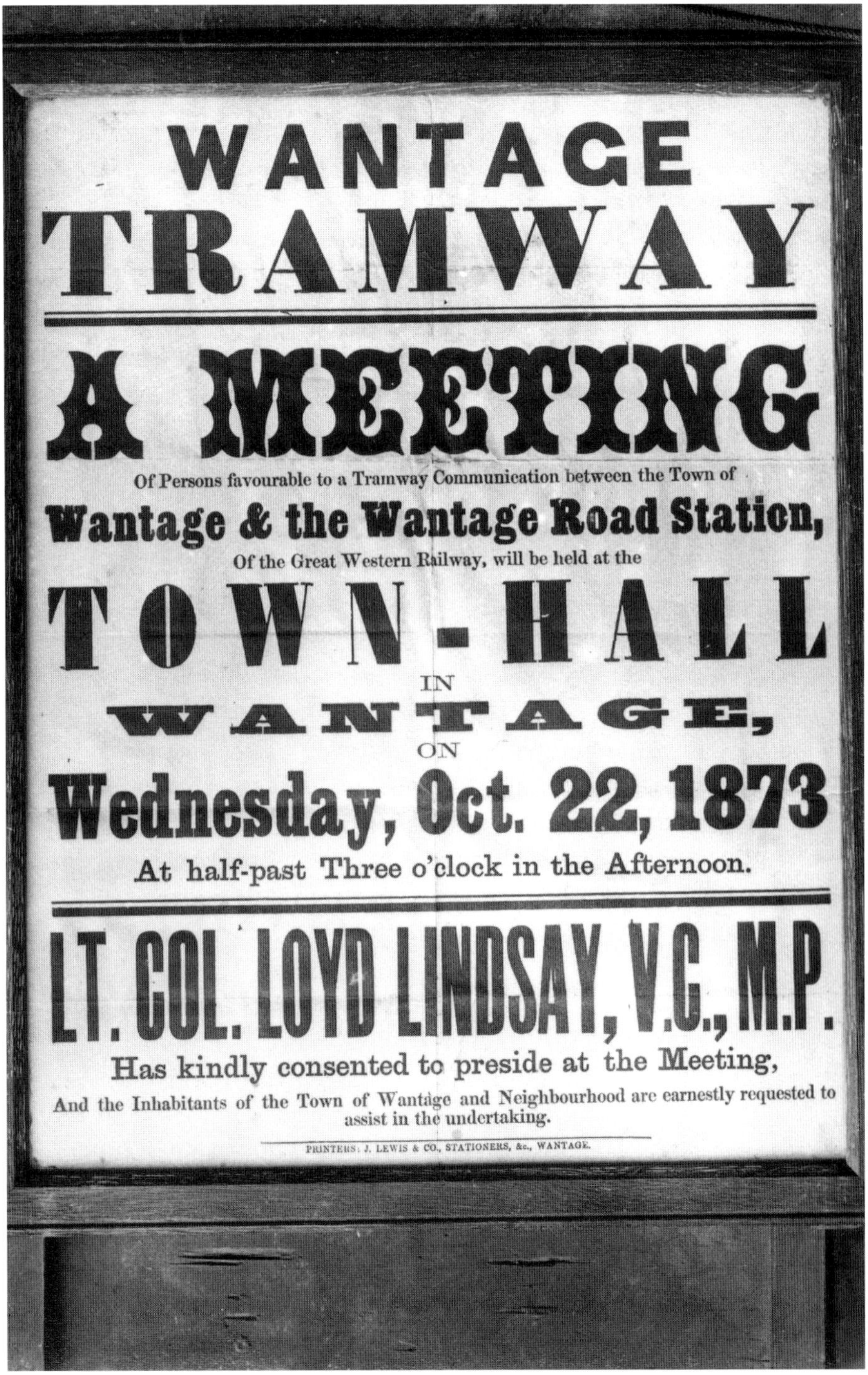

Where it all began. The poster advertising the meeting in Wantage at which the Wantage Tramway Company would be first formed. *Vale and Downland Museum, Wantage*

Ormond's words must be viewed, with the benefit of hindsight, with some degree of scepticism. As a promoter of the tramway scheme, he was likely to suggest that 'his' tramway plan was a better solution to the railway promoter's project. Whilst it is true that a tramway would be considerably cheaper to construct due to the lighter engineering required relative to the building of a railway, the assertion that the tramway would always provide a return on the shareholder's investment seems optimistic, since there was no way to tell, other than in principle, how the traffic on either line would have developed. It would also be fair to assume that the operating costs of the railway, with more expensive rolling stock, would be higher than that of a tramway, but as discussed in Chapter 2, the GWR had agreed to operate the railway on behalf of the junction company, and no doubt would have absorbed the branch, as it did with many others, at a later date. There would be costs associated with operating the tramway which may not have been considered at this point in the company's history. These costs would become clear once the operating agreement with the GWR had been concluded.

Ormond's presentation to the assembled audience moved on to discuss what the promoters had done to prepare for the project to start. The road commissioners who managed the Besselsleigh Turnpike had been consulted and no objection had been received, subject to the tramway collecting the relevant tolls from passengers until 1881 when the jurisdiction of the Turnpike Trust would lapse. The Highway Board had been approached with regard to building the line through Grove and again no objection had been offered. The Great Western Railway had also agreed to the tramway being built into its yard at Wantage Road, subject to a set of conditions being met which would be laid down in a later agreement should the scheme proceed.

The promoter's negotiations with other parties along their proposed route had not proceeded smoothly. Ormond stated that an approach to the Wilts and Berks Canal Company requesting permission to place a bridge across the canal to allow the tramway to cross over it was not met as favourably as elsewhere. The canal company granted permission for the tramway to build their bridge, but at the same time imposed a 'penalty' of £100 simply for its presence. The conflict was eventually resolved, the tramway getting unconditional permission from the canal company to construct the bridge, although it seems that this was given grudgingly.

Stevenson's route was then presented to the meeting with an estimate of the traffic that would be transported along the route each year, based upon the traffic passing along the road. The promoters seemed at this point to have continued to make assumptions that everyone would use the tramway once it had opened. The number of passengers who had booked travel tickets at Wantage Road was stated as having been 8,752 over the previous six months, and the promoters assumed optimistically that all of them had returned and thus that there was a potential annual passenger traffic for the tramway from a total of around 35,000 journeys made from the GWR station each year. The total tonnage of goods passing through the GWR yards was estimated to have been 29,588 tons over the same year, which provided a rich potential source of traffic for the tramway.

The promoters had assumed that the whole of the traffic, both passenger and freight, which had passed through Wantage Road station and on to Wantage town would become traffic on the tramway. This was obviously not going to be the case, with companies such as the Vale of the White Horse Iron works in the town already making their own arrangements, which they continued to do. The promoters suggested to the meeting that they expected the two-thirds of passengers arriving at Wantage Road who had then travelled on to the town using Nunney's Omnibus, would use the quicker tram.

It was suggested that three fifths of the goods traffic arriving at the GWR station was being sent on to Wantage through a variety of means, which could be achieved more quickly and cheaply by the tramway, with three quarters of all the parcels traffic similarly being destined to be delivered within Wantage. The total income, based upon the traffic estimates made by Stevenson, was said to be likely to be £4,164 per year.

Stevenson had estimated that a single line tramway with passing places would cost around £8,000 to build at standard gauge, with operating and maintenance expenses amounting to £1,200 per year. The suggestion that the tramway would be profitable seemed to have captured the imagination of those present. As events would prove, the promoters were mostly correct in the viability of the tramway, even though it could be argued that the assumptions made at the very beginning were optimistic and some good management would be needed at times to keep the wheels turning.

Three resolutions were put to the vote, all of which were carried. The first proposed that the current transport links with the GWR needed to be improved and that this should be done by the laying down of a tramway. The passing of the motion appears to have been the end of the Wantage Junction Railway scheme. It is highly likely that those who had supported the railway project withdrew their subscriptions to the share offering in the railway and transferred their investment commitment to the tramway.

The second resolution proposed that a limited liability company should be formed to construct and then operate the tramway between Wantage and Wantage Road station. The company was to be financed through the distribution of 2,000 shares at £5 each. The motion was passed, and the Wantage Tramway Company effectively came into existence, pending its registration and the receipt of a certificate of incorporation.

The final resolution proposed the creation of a committee to oversee the construction of the tramway. The suggested committee was to be made up of Colonel Robert Loyd Lindsay, as chairman, with Philip Gibbons, Henry P. Gibbons,

Edward Ormond, Walter Barker, Llewellyn Jotcham, Benjamin Castle, Joseph Lewis, Edward Davey, James Clark, John Meadows and Thomas Thosland as the remaining committee members.

The Memorandum and Articles of Association of the tramway company which followed on 7 November 1873 laid out how the company was to be structured. The company was to be funded as had been agreed and then laid down in the BoT order and the Act of Parliament which followed, through 2,000 shares at £5 per share giving the company a possible capitalisation of £10,000. The company would never be fully capitalised, though, with some shares remaining unsubscribed throughout the existence of the tramway. The articles of association laid down how the directors were able to call on the share subscribers for payment of part of the value of their shares. The company was to call off a value set by the directors with twenty-one days' notice, with all subscribers liable to pay the amount called at the time set by the directors. Subscribers who had not paid on the date set were to be liable to an interest charge of 5 per cent per year on the value called until such time as the amount had been paid.

The first directors of the company were nominated in the articles of association, as was a requirement that anyone wishing to be nominated to become a director of the company must have a shareholding in the company of at least twenty shares. The first directors of the company were to be Lieutenant-Colonel Robert Loyd Lindsay (army officer) as chairman, Walter Baker (surgeon), Llewellyn Jotcham (solicitor), Philip and Henry Gibbons (engineers), Benjamin Castle (farmer), Joseph Lewis (chemist), James Clark (corn dealer and miller), Lawrence Surmon (corn dealer), John Meadows (chemist), Edward Davey (wine merchant) and Thomas Tosland (draper). All the directors, with the exception of Benjamin Castle, who lived nearby in Charlton, were resident in Wantage.

The company was registered under the auspices of the Companies' Act (1862) shortly afterwards. The Wantage Tramway Company Limited was incorporated on 10 November 1873 as company No. 7790, a fee of £6 5s being paid to register the company. The company's first registered address was at the offices of Edward Ormond in Wantage. The Wantage Tramway Company had been born.

The company directors convened for the first directors' meeting of the company on 26 November, with John Meadows acting as chairman in the absence of Loyd Lindsay who was unable to attend. The other directors present were Philip and Henry Gibbons, James Clark, Thomas Tosland and Joseph Lewis. At the meeting, Edward Ormond was formally appointed as company secretary and the London and County Banking Company as the company's bankers through their branch in Wantage. The meeting was then ended, and the company had legally begun to trade.

The tramway had meanwhile been in negotiation with the GWR to try to gain their support for the undertaking. Edward Ormond spent a considerable time meeting with Great Western senior management in Paddington, with a huge archive of correspondence surviving between him and the GWR. The Wantage Tramway's directors were keen to ensure that they had a good working relationship with their larger partner before the first rails were laid and to this end passed the draft texts of the intended provisional order application to the railway company for their advice and input. The Great Western's solicitor's office in Abingdon Street, Westminster was involved in the process throughout, as was the GWR's Chairman James Grierson.

The tramway received considerable advice from Paddington and the solicitor's office, including amendments to be made to the order which would be to the advantage of the GWR and the tramway. Paddington advised Ormond that, so long as they received an undertaking from the tramway company to make the advised changes to the provisional order application, they would have no objection to the tramway's application to the BoT and would support the eventual confirming Bill enforcing the order when it was introduced to the House of Commons in due course.

Whilst the tramway had been in discussion with the GWR, the directors had been considering the manner in which the tramway should be built. At the first meeting of the directors on 28 January 1874, the meeting appointed three of the directors to form a committee to examine similar tramways elsewhere in Britain with a view to incorporating the best parts of each into the project to build the line at Wantage. Benjamin Castle and the two engineers Philip and Henry Gibbons who owned and ran The Vale of the White Horse Foundry and Engineering Works in Wantage were appointed to perform the task and to report back the board. Their findings would then be used to instruct the tramway's engineer on best practice from elsewhere.

The next meeting of the directors agreed that the first Annual General Meeting of the company's shareholders should take place at 4pm on Monday, 9 March 1874 in Wantage Town Hall. George Stevenson was directed to prepare a report and specification of work for the tramway based upon Henry and Philip Gibbons's findings and to obtain estimates of the associated costs of that work for presentation to the shareholders at the meeting. In the meantime, P. Gibbons and W. Clark were appointed to define the required buildings for the terminals at each end of the tramway.

The first directors' report presented to the shareholders by Philip Gibbons in the absence of Loyd Lindsay who was detained elsewhere, showed the progress which had been made in the first six months of the company in addition to the registration of the company, and the appointment of directors. The requirements of the BoT had all been met prior to the meeting. All the necessary advertisements had been placed in the county newspapers and the *London Gazette*, with notices advertising the intention to apply for a provisional order to build the tramway posted along the road on the entire route of the proposed tramway with extra posters placed in Mill Street in Wantage in accordance with the requirements of the Tramways Act.

BOARD OF TRADE, Session 1874.

Wantage
TRAMWAYS.

NOTICE IS HEREBY GIVEN,

That application is intended to be made to the Board of Trade, on or before the 23rd day of December next, for a Provisional Order, under the above name or short title, whereby it is proposed (among other purposes) to authorise the making and maintaining of certain Tramways in the Hamlet of Grove in the Parish of Wantage, and in the Town of Wantage, in the County of Berks, and that one of such Tramways is proposed to be laid along the street or public highway in which this Notice is posted.

At no point will such Tramway to be authorised by the proposed Provisional Order be so laid that for a distance of thirty feet or upwards, a less space than ten feet six inches shall intervene between the outside of the footpath on either side of the road and the nearest rail of the Tramway, except in that part of Mill Street, Wantage, which lies to the East of the dwelling house in the occupation of Mr. John Dixon, and extends Eastward for the distance of fifty yards up to its junction with the Market Place of Wantage, and also on the Northern side of the said Market Place for the distance of seven yards measured in an Eastwardly direction.

The Owners and Occupiers of houses, shops, or warehouses abutting upon any part of the street or public highway or place where such less space shall intervene, if they dissent from the Tramway being so laid, may express their dissent by a statement, in writing, addressed to the Assistant Secretary, Railway Department, Board of Trade, Whitehall, London, and made at any time before the 15th day of January, 1874.

Dated this 15th day of November, 1873.

EDWARD ORMOND,

Wantage, Solicitor for the Promoters.

EDWARD WALMISLEY,

25, Abingdon Street, Westminster, Parliamentary Agent.

J. Lewis & Co., Printers, &c., Wantage.

The notification poster which was posted along the line of the proposed tramway in order to meet the requirements under the Tramways Act, 1870, for an application for a Tramway Order. *Vale and Downland Museum, Wantage*

The proof that the advertisements had been placed in the *London Gazette*, the *Reading Mercury* and the *Wantage Chronicle* was to be sent to the BoT prior to the application for the tramway order.

The directors informed the assembled shareholders that George Stevenson had completed his survey and that the plan and section maps of the route had been created. Lithographed copies of them had been created and deposited with the BoT and the Clerk of the Peace for Berkshire, as well as the relevant parish clerks, road and local authorities. The plans and section of the intended route, with the estimates of cost associated with building the line, had been inspected and approved by two men who the directors of the company referred to as 'practical engineers', although unfortunately they did not name the gentlemen concerned.

The company had appointed Edward Walmisley of Abingdon Street, Westminster as their parliamentary agent to act in Westminster on the company's behalf. Walmisley had then lodged further copies of the tramway plans with the Private Bill Office of the House of Commons and the Clerk of the Parliament in the House of Lords. All associated fees had been paid, including £288 deposit paid to the Court of Chancery on behalf of the company by Walmisley on 15 May 1874. The prerequisites had been completed and the way ahead cleared for the company to submit their final application for the BoT provisional tramway order.

The directors had written and signed the required memorial, the wording of which was specified in the 1870 Tramways Act, requesting the Board grant the order to the tramway. The directors had also, as specified in the Tramways Act, written the wording of the order to be requested, and had had a printed draft of it prepared. The documents had been prepared by Edward Ormond and Edward Walmisley, who had both been in frequent correspondence with the Railway Department of the BoT to ensure that no objections were likely to be encountered from the board when the final application was submitted.

The initial application indicating an intent to lodge a request for the provisional order had been submitted to the Railway Department of the BoT on 15 November 1873, paving the way for the formal application for the order to proceed. The memorial, draft wording of the provisional order which was being requested and estimates of the cost of constructing the tramway had been sent to Walmisley, who had deposited the first pack of documents with the railways department of the BoT on 11 February 1874.

No objections had been received during the time since the intent to apply for a provisional order had been advertised from any of the residents or businesses in Grove and Wantage, and consents had been received from the road and local authorities supporting the tramway. The BoT's legal advisors had therefore informed the directors of the company that the provisional order was going to be approved by the Railway Department of the BoT. The speed that this was achieved suggests that the omission of any mention of plans to use steam power may have been a deliberate decision to avoid any potential objections within the Board, although no evidence to support this either way has survived.

Importantly, the shareholders were informed that the chairman Loyd Lindsay and Edward Ormond in his role as company secretary had been in discussion with the directors of the GWR at Paddington. The Great Western had expressed a willingness to assist the tramway, and to allow a junction to be created between its lines and those of the tramway at Wantage Road, as well as the laying of the tramway's lines in the yard. Importantly, the railway company had agreed to appoint the tramway as their agents for passenger, goods and parcels traffic from Wantage Road to the Wantage town district. The GWR was to serve notice on the current holders of the agency agreements in accordance with the clauses laid out in the railway company's agency agreement. The Tramway company did not buy out the businesses who held the rights at the time, the severance was performed by the GWR with a new contract sent to the Wantage Tramway Company once the existing relationships had been severed.

The cost of constructing the tramway had been assessed and estimated at around £1,600 per mile. The directors believed that the costs of the entire tramway, including the purchase of land, construction of stations and other buildings, purchase of rolling stock and all preliminary expenses associated with surveys and legal costs in Westminster, would amount to no more than £7,500.

The tramway's share offer had been well subscribed in its early days with 944 of the £5 shares being subscribed to by fifty members. In order to fund the initial expenses of the tramway, a call would thus be made on all members of the tramway project amounting to 10 per cent of the value of the shares they had subscribed to.

The shareholders were informed, however, that the company would be unable to enter into any contracts related to the construction of the line until the provisional order had been authorised by the BoT and had passed through both the Houses of Commons and the Lords. This was viewed as being a formality and the provisional order authorising the construction of the tramway between Wantage Road and Wantage Town passed in May 1874. The cost associated with achieving the order, whilst considerably less than that associated with meeting the requirements for a full railway order, was still relatively high. The work undertaken by Edward Walmisley as parliamentary agent on behalf of the Wantage Tramway in lobbying the BoT for support and ensuring that all of the Tramway Act's requirements had been met in full had been in progress for some time. Walmisley presented his account of expenses to the company in August 1874. The total expenses accrued for work associated with the granting of the provisional order, undertaken between 10 October 1873 and 8 August 1874 amounted to £286.16s.3d. The work undertaken by Edward Ormond was considerable, with frequent journeys to London to meet with representatives of the Railway Department of the BoT, and large amounts of administrative paperwork to be submitted through Walmisley. Unfortunately, the surviving account books do not itemise the exact nature of the work that Ormond undertook, but it is clear from the correspondence files that, as company secretary, Ormond alongside Walmisley and Loyd Lindsay, had been instrumental in getting the tramway to the stage of gaining the legislation that it needed. The authority granted under the BoT order allowed the company to now begin the process of constructing their new line.

George Stevenson had been instructed to complete the final specification for the tramway line which Edward Ormond was then to forward to a Mr. Tollit, an experienced engineer. Tollit would inspect the specification and certify that it was complete and sufficient for contractors to tender against, receiving the sum of £10.10s for his services and certificate. He suggested a number of amendments to Stevenson's specification which the tramway board agreed to adopt. George Stevenson, attending the directors' meeting, offered to act as consulting engineer in return for a payment of 3 per cent of the cost of building the tramway and connecting it to the GWR at Wantage Road. The directors agreed to accept and appointed Stevenson on those terms.

The company was now able to advertise in various civil engineering journals for contractors to tender for the construction of the tramway. The advertisement attracted a large amount of interest with various parties being sent the plans and the tramway company's terms and conditions against which a contractor would be appointed. A lot of bids were received from across Britain and Ireland from contractors who wished to undertake the work.

Michael Duffy and James Crutchley of Fenchurch Street in London submitted a bid on 14 September 1874. They were willing to undertake the work for £8,200 in total which included £300 for the bridge over the canal and £100 for lever boxes and cranks to operate points on the loops along the line. No completion date was included in the tender. The directors believed that the bid's cost was excessive and rejected it. The company received further bids on the same day, none of which was deemed acceptable.

R.W. Fitzmorris and Company of Birmingham offered to build the tramway for a total of £6,200, including £300 for the canal bridge and £100 for point levers and associated equipment, but the costs were a little high for the directors who rejected it. A higher bid from Haulaye Daudon and Company of Dublin stated a cost of £7,000 including £300 for the canal bridge, £100 for points, and an extra £200 for road diversions, which had not been deemed necessary by the tramway directors, Stevenson, nor by the road authority. Haulaye Daudon added a note to their bid to the effect that they were experienced railway builders who had worked on numerous railways in England, Ireland and America, and as such could promise the tramway company 'satisfaction'. The bid was rejected, once again due to excessive cost. J.M. Wiswell of Dublin, and George Young of Hammersmith both sent in simple one line letters, received on 14 September, offering to build the line for a total of £7,000, and £8,500 respectively. Both were rejected.

John Williams of Nottingham's bid arrived on the following day. He offered to undertake the work to the company specifications for the sum of £6,352, with the tramway to be completed within two months of work starting. Three London-based contractors' bids were received on the same day. John Sykes offered to build the tramway for £6,620, Thomas Wainwright and Company for £6,150, and Allen Vickers offered a total cost of £5,975. All were rejected as was a bid from John Cokes from the Isle of Wight at a cost of £6,620.

George Stevenson submitted two bids of his own to the directors. The first was submitted in partnership with Joseph Rendell from London, for a total cost of £5,626, the second with John Wilkinson of Wantage for a total of £5,350. Stevenson's bids were both rejected on the grounds of cost.

The options presented by these contractors were discussed at the directors' meeting of 16 September with the outcomes described. The tender to build the tramway which the directors believed was the most acceptable came from Philip Ward of Salop Row, Handsworth, Birmingham. Ward had included in his bid descriptions of work he had undertaken on many other tramways, including for Birmingham Corporation. The combination of the contractor's experience and the bid to complete the work at £4,201, the lowest received, won the business. The directors instructed Edward Ormond to write to Ward advising him of the successful outcome of his tender, conditional upon the tramway receiving two acceptable financial guarantees for the completion of the contract and two satisfactory testimonials from organisations for which the contractor had built tramways.

At the same time, the company had been continuing to seek agreement from the GWR on the terms under which the agency agreement would be drawn between the two companies. In order to expedite the conversion, Walter Barker, Benjamin Castle and Philip Gibbons were to meet with James Grierson, the GWR's general manager, in Wantage. Grierson was unable to arrange a trip to Wantage, his work at Paddington preventing him from taking the time out of the office to attend the meeting. Instead, a meeting was arranged at Paddington with the three directors of the tramway to travel to London and meet Grierson in his office. The meeting was attended only by Edward Ormond representing the directors of the company, and George Stevenson as consulting engineer. Grierson assured the Wantage men that the GWR was comfortable with agreeing an agency with the tramway and that he had instructed the solicitors office to draw up the contract on the GWR's standard agency terms.

The Wantage Tramway Company received the guarantees from Philip Ward during September and issued a contract for the construction of the tramway. The contract stated that George Stevenson was appointed engineer to oversee the building of the line, which Philip Ward was to construct for the sum of £4,201. The work had been secured through a joint and several bond provided by Philip Ward and John Oldham of Birmingham. The agreement laid out that the tramway was to be built in 'a good tasking and workmanlike manner' but most importantly 'in the manner expressed or contained in the...plan and specifications'. The latter clause would later prove important when a disagreement arose. The contract specified that the work on the line was to completed by 27 April unless Ward was 'hindered by the shirking of his workmen'. The contract was duly signed by both parties on 29 October 1874 and with the BoT's Provisional Order in place, work could begin on the line.

The directors had not included the construction of the tramway stations in Wantage and at Wantage Road in the tender for the tramway. The facilities at Wantage Road were to be built by the GWR on behalf of the company and so were not available to bid upon. The station at Mill Street in Wantage was, however, to be built by an independent contractor. Local builders were invited to tender during December 1875 for the construction of the buildings, with Whiting & Tosland, William Gregory, and George & John Kent all bidding for the work. The contract was awarded to Whiting and Tosland as the lowest bidders, with all the buildings to be constructed to the

The Matthews tram engine with passenger car 4 at Wantage Road station being loaded with luggage. The steps by which passengers could enter or leave the car are being stood on by the gentleman in the boater hat. The GWR's main line is at the rear of the picture. The condition of the track and the ground in the yard bears comparison to the somewhat less impressive state that the upper yard at Wantage could sometimes be found in. The tram engine's bar coupling is folded away to the left side of the engine. *John Scott-Morgan Collection*

drawings provided by George Stevenson. It is notable that all the builders bidding for the work were also shareholders of the tramway scheme.

The Provisional BoT order authorising construction of the tramway was converted into a formal Act of Parliament in 1874, passing on 7 August. The Wantage Tramway was included with a number of other projects which had received provisional orders in the Tramway Orders Confirmation Act, 1874. It included confirmation of authorities under provisional orders for Birmingham (Corporation) Tramways to amend their 1872 order; London Street Tramways for the construction of additional tramways in St. Pancras and Islington; The Newbury and Lambourn Tramway which would later become the Lambourn Valley Railway; Portsmouth Street Tramways for additional tramways in Portsea; The Wantage Tramway for the construction of the complete tramway; and finally Wirral Tramways for the construction of new tramways in Birkenhead and Tranmere.

The Wantage tramway was split into separate sections as was usual in railway and tramway Acts. The main tram line between Wantage Road and Wantage was described as Tramway No.1. The route was laid out in the Act as starting in the GWR's yard at Wantage Road and then proceeding out of the yard and on towards Wantage. The Act is unusual in specifying that the tramway was to be built along the east side of the Besselsleigh Road, rather than the usual assumption that it would be built along the middle of the road. This is the first indication that the directors were planning from the outset to use steam traction on the tramway and so elected to have the line built not in the road but alongside it. Effectively the Wantage Tramway was about to invent the Tram-Train so common on urban transit tramways in the 21st Century.

The tramway was to then cross the Wilts and Berks canal using its own bridge next to the road bridge over the canal. Once over the canal, the tramway was to cross the corner of a field before returning to the side of the Besselsleigh Road which it would follow until crossing it and diverging away to the south on an incline. The line continued through tramway owned land, the meadows belonging to Edward Ormond and James Palmer having already been sold to the company. The route then ran along the centre of a private occupation road belonging to Wantage Gas Works, and across the corner of an orchard which the company would purchase from George Curtiss. The orchard would later provide a small income for the tramway as part of it was leased, whilst the remainder provided the company with a crop of apples each year to be sold. Once past the orchard, the line was to continue through land owned by the tramway company at the side of Mill Street, then crossing to the north side of the road and entering Market Place, terminating at a point only 4 feet from the corner of Wantage Town Hall. This latter part of the tramway was never constructed, the company opting to terminate their line at Mill Street.

The Act included five additional 'lines' which the tramway company had originally been authorised to construct by the provisional order. Each was a loop 1

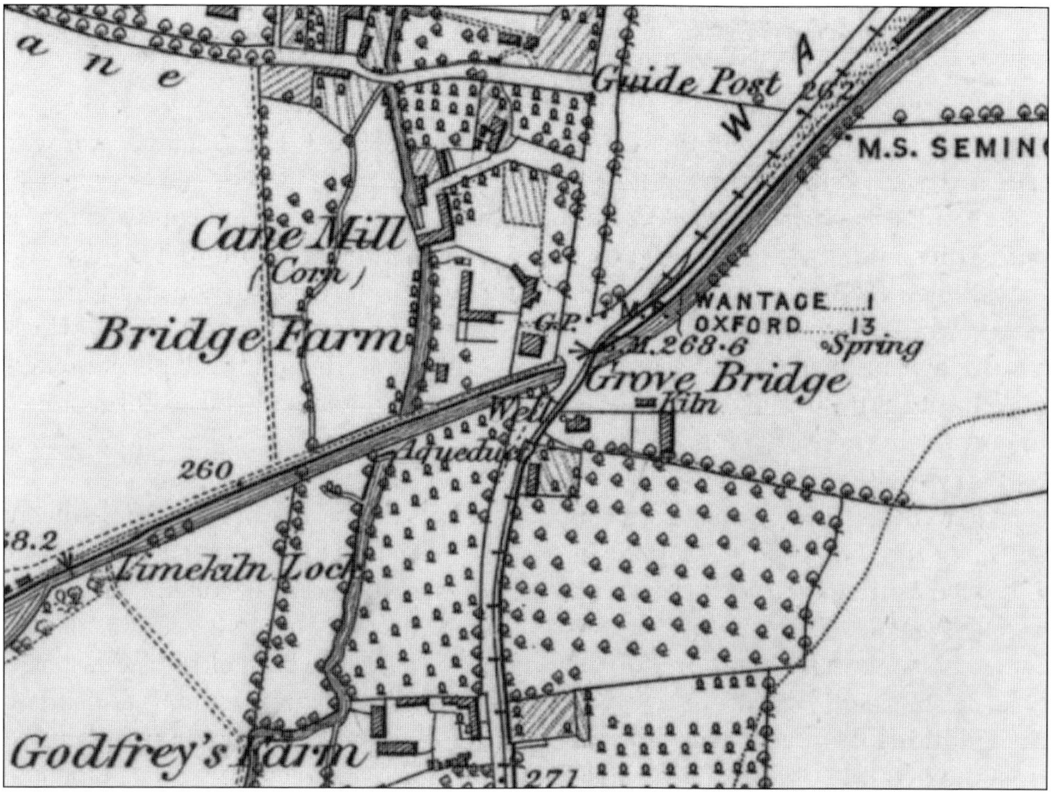

Grove canal bridge and surrounding area showing the location of the tramway line. *Excerpt of Ordnance Survey Six-Inch Series, Berkshire Sheet XIV, 1883. Reproduced with the permission of the National Library of Scotland*

chain and 25 links in length (165ft). The first was to be built at Oxford Lane to allow trams to pass, and a second at Grove on the edge of the footpath leading to Charlton. The third loop was to be located just after the bridge which crossed the Wilts and Berks canal, with a fourth in Grove near the end of the private road leading off the Besselsleigh Road into Loyd Lindsay's estate. The final loop authorised was to be in Edward Ormond's meadow. Not all of the loops were built.

The tramways, and Wantage was no exception, were granted an exemption under the Tramways Act which the railways, as common carriers, were not allowed. They had the option to refuse to carry any item or animal which they thought was unsuitable, or that they believed they were unable to economically transport. The tramway was however to transport the personal luggage of each passenger up to a total weight of 56lb in first class and 28lb in second class without refusal or additional charges over the cost of the passenger ticket.

The authorising Act contained one final clause which determined the relationship between the tramway company and the GWR. The clause stated that the tramway

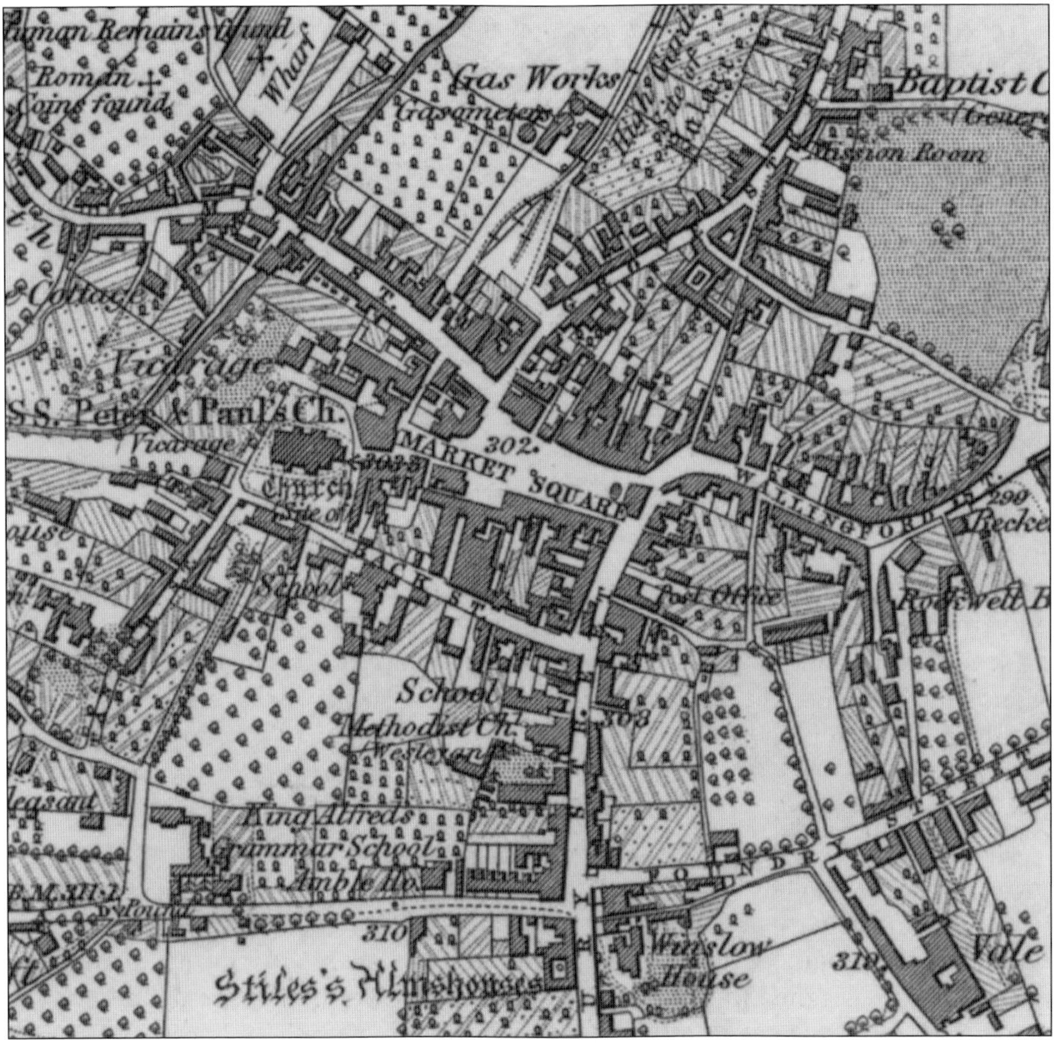

Wantage Town Centre. The tramway terminus is in the top centre of the map near the gasworks. The large building at the lower right corner of the map is the Vale of the White Horse Iron Works. *Excerpt of Ordnance Survey Six-Inch Series, Berkshire Sheet XIV, 1883. Reproduced with the permission of the National Library of Scotland*

was not to construct any part of the tramway within the yard at Wantage Road station without the authority of the Great Western. The GWR was given authority to perform any work on behalf of the tramway, but at the tramway company's cost, related to the laying down of the tramway and its buildings, as well as any changes or improvements which may later be required. The tramway works at Wantage Road were, even with the agreement of the GWR, to be undertaken under the supervision and to the approval of the Great Western's engineer.

The provisional order did not mention the use of mechanical haulage on the tramway, the default clauses of the Tramways Act assumed the use of animal haulage unless otherwise specified. The Act of Parliament which followed thus made the same assumption and precluded the use of steam or other mechanical means of propulsion of the Wantage Tramway. Despite the lack of permissions in the order and its superseding act, the tramway company directors intended to use steam power at the earliest opportunity.

The tramway was only officially able to use horse power on the line and whether this was an oversight by the directors or a deliberate action in order to get the authorities needed to build the tramway is unclear, no mention of it being made in the directors' minute books for the early days of the company. It seems possible that the directors were navigating their way through the process as the BoT were reluctant to allow the use of steam in the early days of the Tramways Act. The BoT appear to have been intending to lobby parliament for an amendment or extension Act to which would set the rules under which steam engines may be used on tramways, and thus did not want to set a legal precedent which may limit their powers under a new act.

Nevertheless, the company would need to gain permission from the BoT before steam traction could be used and in order to do so, an amendment order was required. The company continued to correspond with the board who became more willing to allow the use of mechanical haulage but in the meantime horse power would have to suffice.

The GWR had informed the tramway company that they had no objection to their contractor laying down temporary rails in the yard at Wantage Road in order to allow him to get his materials from the yard onto the tramway as it was being built. The area engineer had been instructed to allow this to be done. Meanwhile, the contractor assigned the task of building the tramway, Philip Ward, had arrived in Wantage with his construction workforce. Raw materials had been arriving at Wantage Road yard on the GWR and by November 1874 with everything that was needed now in place, construction began.

Chapter 4

BUILDING AND OPENING THE TRAMWAY

One of the first tasks that the company had to undertake, before the first shovel hit the ground, was to establish their rights to transport both passengers and goods from Wantage Road into Wantage. In order to do so, Edward Ormond had been discussing an agreement with the GWR at Paddington and was often in correspondence with James Grierson. Grierson seems to have been generally supportive of the scheme to build the Wantage Tramway and often offered advice to the directors of the tramway company. He was also keen to put in place an agency agreement between the GWR and the tramway, although it would mainly be on the GWR's terms.

Whilst discussion had been underway with the GWR, the directors had still been pursuing their ambition to use steam locomotives on the tramway which would result in the company initially taking on trial the Grantham Steam tramcar. The Grantham Patent Steam Tramcar had been originally seen in London by Edward Ormond and George Stevenson in September 1874. The prototype vehicle had been inspected in the District Railway Company's works in West Brompton during the afternoon of 23 September. Despite not having seen the Grantham Tramcar working on the District Railway or on the London Tramways, the men had viewed the Grantham Car as suitable for use at Wantage. Stevenson had been instructed to discuss with the representative of the owners of the steam tramcar, referred to only as Mr. Baxter in the correspondence, the possibility of bringing the steam tramcar to Wantage on hire for a fixed period with a view to purchasing it if it was successful.

Little progress was made, Stevenson not having obtained the costs for hiring the Grantham tram for use on the tramway from the owners. He was instructed to pursue the attempt again to hire the Grantham Tram by the directors of the company after the 14 April 1875 directors' meeting. At the same time, he was authorised to procure the traditional tramcars that would be needed to start the passenger service once the tramway opened to traffic. The new tramcars were to be small carriages capable of carrying a total of fourteen passengers, with the compartment divided to provide accommodation for six first class and eight second class passengers. The first car was

A poor quality photograph of the Grantham Patent Steam Tram, but since very few images of the original car remain in existence it has been included. The open driving positions at each end of the car can be clearly seen. The ladders at each end lead to the open deck where passengers shared space with the engine chimney. The louvred doors on the side of the vehicle led to the boiler on this side of the car, the same arrangement provided on the opposite side. *John Scott-Morgan Collection*

ordered from the Starbuck Car and Wagon Company in Birkenhead, an experienced builder of horse drawn tramcars, at a cost of £180. The company's plans seemed to be progressing well, but a potential problem was about to confront the company directors.

The relationship between the GWR and the Wantage Tramway that would serve the tramway so well throughout its existence was nearly soured at the first step. The railway informed the tramway company that the plans which had been submitted to the area engineer had not included any drawings of the proposed shed that was to be constructed on the tramway's behalf by the GWR. The company viewed the passenger shed as necessary and asked for the plans to be submitted at the tramway company's earliest convenience. A drawing was made and sent to the area engineer shortly thereafter.

The railway company had in return given the tramway an estimate of the cost of undertaking all the works, including the laying of the rails and constructing the tram shed, in Wantage Road station yard. James Grierson had written in a letter, dated 19 May, that the GWR's engineers had examined the plans and visited Wantage

Road to assess the work that would be needed to complete the work. The estimated costs supplied to Grierson were £714.

George Stevenson had written to the GWR informing them that he thought the charges were excessive, the General Manager's Office at Paddington responded in a letter of 22 May 1875 to Edward Ormond in which James Grierson invited Ormond and Stevenson to come up to Paddington to discuss the plans and costs with the GWR's senior engineers. The disagreement was resolved during the meeting that followed with the GWR's fees being accepted, but with the tramway electing to only have part of the work completed. The remainder of the planned facilities at Wantage Road would be done as traffic increased and the levels justified the extra expenditure.

As a result of the decision not to construct all of the tramway's facilities at Wantage Road which had been proposed by Stevenson's drawings, the GWR revisited the works to be undertaken. The tramway and the railway arranged that one of the planned sidings would not be built at all, with the arrangement of the remaining sidings and passenger terminal being rationalised and modified by the GWR's engineers to make improvements on the general layout of the terminus. The Great Western proposed a revised charge of £550 to undertake the revised arrangements, which the tramway accepted. James Grierson in correspondence with Edward Ormond, suggested that if the tramway could place the funds on account with the GWR, he would instruct work to be started immediately the cheque was received.

The two companies had also come to an arrangement with regard to the traffic from Wantage Road to Wantage itself. In accordance with verbal agreements reached between Edward Ormond representing the Wantage Tramway Company Limited and James Grierson as General Manager of the GWR, the railway company proceeded to inform both James Sims and Thomas Nunney that their agency agreements would be terminated. The Great Western wrote to both men informing them that the railway was serving notice in accordance with the clauses of their agency contracts with the GWR that the relationship between them would cease.

The railway company informed the tramway a few days later that they had not had any response from either Nunney or Sims but nevertheless the agency agreements had been terminated correctly and a new agency would be assigned to the Wantage Tramway Company in due course. In a letter to Edward Ormond from the Goods Manager's Office at Paddington, dated 27 September, the tramway company directors were asked under what terms the tramway would be willing to collect and deliver both goods and parcels within a one mile radius of the centre of Wantage Town station once it were opened.

In the same letter, the charges for the use of GWR stock over the tramway were laid out. Despite the tramway moving the GWR's goods wagons over the line using their own locomotives and horses in order to fulfil the Great Western's delivery contracts with its customers and the collection of goods in loaded wagons for onward transport by the GWR for which they were charging a carriage toll, the railway was to charge a rental for the freight stock whilst it was on the tramway. The charge for wagons was to be 1s for the first day or part of a day, with an additional charge of 3s per day after the first day. An additional charge was to be levied on the tramway for the use of wagon sheets (tarpaulins to cover goods in open wagons) at the rate of 6d per day after the first 72 hours if the wagon was loaded in each direction, or after 48 hours if the wagon was only loaded in one direction.

The tramway and the Great Western were still in parallel discussions about the construction of the terminus at Wantage Road. The two companies had agreed the sum that the tramway would pay to Paddington for the GWR to build the facilities for them. In a letter to Edward Ormond, dated 29 June 1875, James Grierson acknowledged receipt of the tramway company's cheque for £550. Grierson advised that he had issued instructions to the GWR's engineering department for the purchase of the needed materials immediately so that no time would be lost in the commencement of the works on the tramway's behalf. The tramway was also informed that the GWR would undertake the laying down of the lines, sidings and terminus buildings at Wantage Road at cost, no profit was to be made. Once the works had been completed to the satisfaction of the both the GWR's engineer and George Stevenson, any balance remaining from the £550 paid on account would immediately be returned to the tramway company.

The final barrier in the way of the tramway company's plans had been the disagreement with the Wilts and Berks Canal Company. The tramway needed the canal company to allow them the right to build a bridge across the waterway but had not been able to do so. The canal company were insisting on a large payment for the right of way needed to build the tramlines across the canal, and this was finally given when the directors of the tramway agreed to the demands of the directors of the canal and paid them £100 for the rights.

Whilst the plans for work at Wantage Road were being discussed with the GWR at Paddington, the tramway had been searching for suitable large passenger tramcars to complement the small cars that had already been ordered. George Stevenson had been visiting a variety of tramcar builders to assess their offerings, with a view to reporting back to the board about their suitability. Walter Barker, James Clark and Thomas Tosland had been appointed as the 'tramcar committee' to assess the drawings and specifications of the tramcar that Stevenson had inspected before authorising the purchase of one additional car.

The company directors, conscious of the ongoing work to build the tramway line, agreed that decisions needed to be made quickly to resolve any eventuality that occurred during the ongoing construction works. Having to wait for the next directors' meeting was obviously going to cause delay, and so the directors agreed to delegate authority to a construction committee who would be authorised to approve any urgent work that was needed prior to opening the tramway. Barker, Clark and Tosland were appointed to a 'works committee' with the authority needed to keep construction progressing, and to ensure that the project delivery dates were met on time.

An agency agreement with the GWR was still proving troublesome to get. The GWR's Goods Manager's Office wrote to Edward Ormond on 30 September with an explanation of the difficulties which were causing delays from Paddington. The agreements that were in place with Sims and Nunney needed a three month notice of cancellation, which had now been given, and until then the railway had been unable to look upon the tramway company as their agent. The GWR had received a response acknowledging the termination of his agreement from Nunney but Sims had not responded. Despite Sims's refusal to acknowledge the loss of his agency agreement, the GWR had instructed Richard Stonhill, the station master at Wantage Road to hand all goods and parcels to the tramway once it had opened, in anticipation of a formal agency agreement between the two companies that would follow.

The Wantage Tramway Company, as a result of the impending reassignment of the GWR agency and to avoid competition, made their own separate agreement with James Sims, dated 22 September 1875. The agreement stated that Sims was to give up his agency with the GWR as of 1 October 1875. Sims was, as a result, to cease working for hire between Wantage and Wantage Road station. In return, the tramway company would appoint Sims as their own agent under new terms. He was to deliver, using his own horses and carriages, all tonnage (heavy) goods and parcels from the tramway station in Wantage town to the consignees in the district of Wantage. Sims was to collect heavy goods for delivery to the tram station for onward carriage by tram and then the GWR.

Under the terms of the contract, Sims was to allow his offices in Grove Street, Wantage, to be used as the tramway's parcels office. He was to collect all carriage charges on behalf of the tramway company monthly for all tonnage and mileage goods and for all parcels, sending updated accounts and monies on to the tramway company each week. In return, Sims would be paid an annual salary of £220 and be employed officially as the tramway's goods manager.

The search for a second larger tramcar had been ongoing, and Stevenson had found a variety of options which he thought would be suitable. He had inspected one car in Birmingham which was complete and ready to be purchased. Returning to the builder's workshops for a second inspection accompanied by James Clark it was agreed that the company would offer £150 for the car, which was to include delivery to Wantage Road station. If this offer was not accepted, Stevenson was authorised to approach the Starbuck Car and Wagon Company, already constructing the first tramcar for the company, with a view to ordering a second, larger, car. The Starbuck

Car and Wagon Company was later renamed G.F. Milnes & Company in 1886 and would go on to build electric tramcars for the Manx Electric Railway.

Construction of the tramway lines had been proceeding and in accordance with the tender let by the Wantage Tramway Company, Ward invoiced the company in instalments as work was completed and signed off by George Stevenson. At first, work progressed well, with Ward presenting his requests for payment on 9 December 1874 for £400, 15 January 1875 for a further £400 and 6 February for £500. Construction of the tramway then encountered some problems, with Ward presenting invoices for additional remedial work at £35 and £30 on 6 and 24 February respectively. Further additional bills for work outside of the agreed contract were presented on 19 March (£50), 17 April (£62.2s.), 25 June (£20.12s) and 8 September (£30). All were accepted by the company and paid.

Despite the out of contract works, construction of the tramway continued to proceed steadily, with Ward submitting requests for payment on 6 March (£500), 7 April (£500), 28 May (£600), 7 August (£400) and finally on 10 December 1875, a bill for £332 18s 5d was presented to the company, completing the work. Once construction had been completed, the tramway company purchased all of the remaining materials located at Wantage Road station that the contractor had not used, a further payment of just over £70 being made for them.

During the construction of the tramway, the company found that it was short of capital, mainly due to some of the shares in the tramway remaining unsold. In order to address the cash flow problem, Edward Ormond approached S.M. Morland with a view to seeking assistance from Lord Overstone. Overstone was the father of Harriet Loyd, the wife of Colonel Robert Loyd Lindsay, and was supportive of the tramway scheme. Morland wrote to Ormond on 8 June 1875 confirming that Lord Overstone was prepared to advance a loan of up to £3,000 to the tramway company. The company took out a loan of £1,800 on 23 June 1875, adding to the eighty shares (£400) that Overstone had already purchased in the company. The additional funds injected into the company took the investment up to near the £10,000 that had been anticipated to be needed and allowed the building of the tramway and the purchase of materials and equipment needed to open the line to proceed.

As construction was proceeding well, the tramway company directors began to look towards the opening of the line and the horses that would be needed to haul the traffic. In order to quickly put in place the necessary arrangements, rather than buy horses and arrange for their stabling and care, it was decided that the provision of the animals should be subcontracted to a local company. As a gesture of goodwill following his loss of the passenger agency with the GWR, the tramway company directors approached Thomas Nunney to establish if he would be prepared to supply his horses on a contract basis. Nunney agreed and the contract was signed on 14 July.

The Grantham Steam Tramcar had arrived at Wantage during July 1875, with arrangements for its hire and possible purchase having been agreed with the owner's representative. Although it had arrived, the directors of the company issued an instruction following the director's meeting of 14 July that the Grantham Tram was not to be used on the company's rails. This was not through any reticence to use the machine, or because the tram had been brought to Wantage by Colonel Loyd Lindsay. The directors, as has already been described, had been involved in the inspection of the steam tram and had been keen for Stevenson to arrange for its hire and trial with a view to acquiring the machine. The instruction that the Grantham Tram was not to be used was driven by three other factors.

Firstly the company did not have authority to use steam on the tramway when it opened, the tramway order specifying that animal power was to be used, and although an application was in the process of being made for an extension order allowing the use of steam, it had yet to be granted. Secondly the tramway simply did not have a driver who knew how to operate a steam locomotive of any kind and until that person was appointed the BoT was highly likely to take a dim view of the operation of the vehicle, which may well have jeopardised the granting of the steam order and ruined the good relationship that the tramway company had with the railway department of the BoT. The tramway did however write to the board requesting temporary permission to use the Grantham car pending the granting of the steam order extension, which the BoT unofficially granted in due course, but with a caveat that operation of the Grantham tram was to be undertaken at the risk of the company alone.

Finally, the company had not been able to reach an agreement with Baxter and the Grantham Tram's owners with regard to the cost of using the vehicle and until such time as that had been settled, it was commercially inadvisable for the tramway to use the vehicle as the owners could potentially impose a hire fee upon the company without the tramway having much chance for negotiation.

The station buildings at both Wantage Town and Wantage Road had not been part of the contract for the construction of the line, as outlined in the previous chapter. Although the terminal buildings at Wantage Road were to be built by the GWR on behalf of the tramway, the directors of the tramway company remained committed to placing the work for the terminal in Wantage with local builders, providing work for the town. Tenders for the work were sought and on 21 July 1875 the contract for Wantage Town station was awarded to Wheeler & Gregory a local firm of builders. The station was completed quickly despite a change to the design of the buildings during the construction works. The passenger platform was to be covered by a roof which had not initially been considered in the design. Wheeler and Gregory provided a galvanised corrugated steel roof which covered 50ft of the platform length, and was 19ft wide, covering both the platform and the tram line ensuring

A view of the passenger terminal in Wantage in the early 1920s with the later pitched roof. The Matthews Tram, No. 6, is unusually sandwiched between passenger cars 3 and 6. The trams could not operate in push-pull mode, so it is likely that the car nearest the camera is the car the Matthews tram arrived with, the other not being used for that service. The brick building at the end of the yard past the trams, is the company's office, parcels office and waiting room. *John Scott-Morgan Collection*

that passengers were completely protected from inclement weather whilst boarding or leaving the tram cars.

The tramway was completed in late August and the directors invited the BoT-nominated inspector, Colonel Hutchinson, to visit the line to perform the required inspection before opening. The directors travelled to Wantage Road to meet Hutchinson using tramcar No. 1, which was horse drawn. The Colonel inspected the line and its facilities, spending additional time performing a detailed survey of the canal bridge. At the end of the visit he congratulated the directors on the construction of the line, although there were some outstanding requirements which needed to be met. He expressed an opinion that, despite the remedial work required, he thought that the line was ready to open for traffic and that he would report to the BoT accordingly.

The construction of the line, despite Colonel Hutchinson's impressions, had not, however, proceeded entirely smoothly. A dispute had arisen between Philip Ward and George Stevenson regarding how the line should be constructed. Stevenson had regularly attended the sites where the line was being built and had instructed Ward's workforce to deviate from the plans agreed in the tender, and also not to lay the line in the manner which Ward had determined and used before on other tramways. Ward approached the board of directors in early November 1875 and asked that he be allowed to attend the next board meeting on the evening of 9 November with Stevenson in order '…for both to be questioned and answered to see who is in the fault'. The board agreed to Ward's request and summoned Stevenson to also attend the meeting.

Ward presented his case to the board. The disagreement had arisen when Stevenson deviated from the original plans for the tramway which had been approved by the BoT. The plans had specified that the line should be laid upon a 6in layer of crushed chalk topped with a further layer of 6in of good quality stone quarried nearby in Marcham near Abingdon. Ward had contracted to deliver this as a good quality bed for the railway to be laid upon. The dispute had begun when Stevenson deviated from the agreed plans.

Stevenson had insisted that the stone, which Ward had already purchased, was not needed and that a simple bed of 6in of poor quality Chain Hill chalk was sufficient for the tramway. Ward stated that, in his experience, and that of other light railway and tramway contractors, a base of the kind being advocated by Stevenson was only suitable for topping the infill of an excavation of up to 10 or 12ft depth. It was unsuitable for a line being constructed adjacent to a Turnpike Road. The Chain Hill chalk was also more expensive, despite being of lower quality than Marcham Stone, costs of around £300 being involved for a poorer quality railroad bed than the £35 that a good quality stone one would cost.

Ward also stated that he was unhappy with the quality of materials being used for the railway itself. The plans had been for the line to be supported upon longitudinal sleepers with cross bracing sleepers at regular intervals to preserve the gauge, similar to the method that had been used for the construction of the Broad Gauge on the GWR by Brunel. Whilst, according to Ward, this was perfectly acceptable and had been used on other tramways, the materials which the tramway's engineer had purchased were of very poor quality, many of the longitudinal sleepers being found to already be rotten when laid down.

Ward had found that when his men had been laying the single 6in layer of Chain Hill chalk as instructed, the chalk was dissolving and forming a thick whitewash-like liquid which had splashed up in the workmen's faces as they tried to lay the line. Ward stated that he had intervened and the tramway had been re-laid according the original specification where construction had deviated. Once completed, it had been found that when 130 tons had been drawn over each stretch, the line was solid with no sinking being observed. Stevenson had then approved the work.

The company's engineer had, according to Ward's testimony to the board, intervened once more. As originally instructed, the contractor had filled the space between the longitudinals with a 6in layer of chalk, bringing the level of the fill to the top of the rails. Stevenson then decided to have the level of the fill between the rails raised by a further 6 inches, using Marcham stone topped with a layer of macadam, but when this was being completed had the stones raked off the tar and a new surface put down. Ward described the replacement surface as being 'like a gravel walk', which was uneven and unsuitable for a tramway. The stones which had been raked off were to be broken up into smaller pieces and laid down on top of the gravel once that had been rolled to compact it. Ward complained that this was both inappropriate for the tramway and had also cost him money, which he had invoiced the tramway for.

Ward informed the directors that the company was also in breach of contract due to the instruction from the engineer that an additional turnout be laid on the tramway and a siding be constructed from it into the orchard 'between Mrs Liddiard's and the ten posts' which had added to the costs. Ward stated that according to the rules of the BoT, the contract could be referred to the Board for arbitration if needed, as he felt the tramway was in breach of contract.

Ward also had issues with the construction work that had been undertaken for the canal bridge. Ward had not received a detailed specification of the works required, although Stevenson had supervised the works himself. Arriving at the canal one morning, according to Ward, the engineer set out the bridge works and the contractor's workers were instructed to follow the instructions as defined for the excavation of the foundations for the abutments of the bridge.

The following day, as the bricklayers were about to begin building the abutments upon the prepared foundations, Ward claimed that Stevenson had arrived and gave instructions that the foundations were to be changed. This apparently happened seven times, with the abutments becoming increasingly larger, eventually being reported to be around 12in above the level of the line once the bridge had been completed. The line needed raising to the level of the bridge with additional ballast. Ward complained that the original allowance agreed in the contract was for a sum of £300 for the construction of the bridge, but the changes had resulted in a cost of £430 18s ½d being incurred by the contractor.

The engineer had duly received invoices for all of the additional works which Ward trusted the company would honour. Ward was also concerned as to the standard of the line as it had been completed. Stevenson was said to have been 'about the line' every day instructing the builders, sometimes wanting the line to be raised above the level of the road it ran next to, sometimes below the level of the road. Ward's concerns were that, in his opinion after undertaking an inspection of the completed works, the tramway had been completed above the road where it should be below it and below the road where it ought to be above it. Water had been observed collecting in these places where the surface between the rails had become boggy. As a result, Ward said that he was willing to put the contract into measure and value with the Board of Trade and for payment to be determined by arbitration. The directors of the tramway replied that this would not be necessary and all of Ward's invoices would be paid in full. They also thanked him for his frankness and advice.

The dispute between Ward and Stevenson is difficult to fathom. Whether Ward or Stevenson was right remains a subject for debate, although it is probable that to one extent or another they were both doing what they thought was best for the line and were trying to build as good a tramway that they could.

A letter from the BoT, dated 30 August 1875, was received by Edward Ormond in his role as Wantage Tramway Company Secretary. The letter stated that the Board had received Colonel Hutchinson's report relating to his inspection of the completed tramway, and that he had found no reasons to delay the opening of the line, having found it in most respects in compliance with the standards laid out within the Tramways Act and thus fit for public use. As such, the BoT was pleased to authorise the tramway to open, and enclosed the licence required for the tramway company to run trains on the line. The licence was provided subject to the tramway company undertaking the recommendations made in Colonel Hutchinson's report of 27 August which they had also enclosed.

The BoT certificate authorising the line to open allowed the company to begin operating trains immediately, however it was decided that opening the line to public traffic should be delayed until after the September directors' meeting so that the remedial works could be discussed and arranged. The 1 September meeting saw the directors agree to undertake all of Colonel Hutchinson's recommendations forthwith, and also to open the line. Goods traffic would be the first to be operated, although using road transport until the most urgent works had been completed. Once done, the passenger service would be advertised and start shortly afterwards, with goods traffic moving to rail shortly afterwards.

Hutchinson's report was mainly complimentary of the workmanship evident in the construction of the tramway, although the Colonel had a number of recommendations which he required to be met before reinspecting the tramway to confirm the licences from the BoT. Hutchinson had found that some of the crossing places had 'been badly laid in' and that they should be put into a proper state of repair immediately. The level crossing which took the tramway across Besselsleigh Road was deemed to have been poorly designed, the gravel path between the rails being deemed insufficient. The crossing required remedial work involving the replacement of the gravel and macadam path with stone pitching between the rails, with 18in of stone setts to be placed either side of the line on the outside of each rail.

The level crossing at the entrance to the Wantage Tramway Company's private land in November 1930. The crossing had been covered with tarmacadam by then, replacing the stone sets which Colonel Hutchinson had insisted upon. The line which carries on straight into the distance leads to the original upper yard, whilst the line branching off to the right heads into the lower yard opened in 1905. *John Scott-Morgan Collection*

The Colonel's inspection of the tramcar which had been purchased for use on the tramway found that the manufacturer's design was not suitable for use on the Wantage Tramway due to the difference in construction from an urban street tramway. Hutchinson required that the wheelsets were replaced with wheels having a deeper flange depth, more akin to that used on railways than urban tramways, to prevent the cars from derailing. Once the remedial work had been completed, Hutchinson suggested that the tramway contact him to arrange for a short visit to certify the work so that the BoT could remove the conditional permissions of the licence.

Hutchinson's report provides a valuable glimpse into the completed new tramway. He found it noteworthy that rather than use the usual U section grooved rail commonly found on urban tramways, the Wantage Tramway had been built using bridge rail of the type used by the GWR on their broad gauge. The rail looked similar to the later flat bottomed rail still used on railways across the world in the twenty-first century, but the rail itself was hollow at the bottom, forming the shape of an arched bridge in section. This is understandable given that the rails had been purchased as second hand stock from the GWR and were still more than serviceable for a low speed tramway.

The line had not been filled with stone paving between the rails, instead a macadam and gravel path had been used, which Hutchinson reported neither the trustees of the Besselsleigh Turnpike Road nor the local road authority had objected to. Other than the level crossing, which had been singled out for improvement, Hutchinson was content with the standard of the macadam surface. The decision to use a road surface between the rails would later cause the tramway company a problem which would contribute to the closure of the line, but in the nineteenth century no such threat could present itself.

Hutchinson, whilst happy with the construction of the 37ft long wrought iron skew bridge across the Wilts and Berks Canal in regard to its strength in relation to the use of horse drawn trams, viewed it as a clumsy structure. He was concerned about the intended use of steam traction across the bridge and required that if the tramway were to pursue their plans to introduce steam locomotion to their line, a reinspection would be required to pass the bridge safe for the increased weight of the steam engines to be used.

Whilst the construction of the tramway had been proceeding, the tramway directors, and especially Edward Ormond, had been continuing to negotiate with the GWR in order to gain an agreement for the operation of traffic on their behalf. A final agency contract was agreed which would come into force on 1 October 1875. The agency contract was quite specific on the expectations that the GWR had of the tramway company. The Wantage Tramway Company was not to undertake any works in the yard at Wantage Road without the consent of the Great Western, and all movements of locomotives, trains, or shunting operations in the yard were to be performed in accordance with the rules, regulations and byelaws of the GWR. The tramway was to be charged a rent of five shillings per year for the right to access the GWR's yard at Wantage Road.

The tramway was now ready to open, although later than anticipated and with some conditions. The company began its services by initially only carrying goods traffic, starting from 1 October 1875 but using road transport. Passenger services began on 11 October using a road omnibus initially, but this was quickly followed by the use of tramcar No. 1, which had been fitted with new wheelsets, hauled by horses along the tramway.

The 5 October directors' meeting confirmed that an order should be placed with the Starbuck Car and Wagon Company for tramcar No. 2 at a cost of £167 10s with an additional charge of £5 7s 3d for carriage to Wantage Road station yard from Birkenhead. It was also agreed, and this seems to be the source of the confusion regarding Loyd Lindsay's involvement with the Grantham Patent Steam Tramcar, that limited trials and experiments with the steam tram would be undertaken under his supervision within the private part of the tramway in the yards at Wantage Town and the short section of line as far as the main road. As this part of the line was on private property and not running alongside the public road or into the GWR yards, the BoT was content to allow the tramway to test the steam tram to establish its capabilities.

The BoT sent a letter to the Wantage Tramway Company on 12 October chasing them for confirmation that the corrective work required by Colonel Hutchinson had been completed and that a final inspection had been arranged with the Colonel. The BoT sent another communication on 18 October in which they responded to the proposed changes to the company byelaws which the tramway company had submitted to the board on 8 October.

The new byelaws were accepted by the BoT with an additional change from the board regarding Byelaw 2, by which access onto and from the tramcars was laid down. The amendment set out that passengers were only to be allowed to enter carriages through the rearmost platform from the direction of travel, as was normal on other tramways, the front platform only being used by passengers alighting from the car. The Board stated that they had no objections to the proposed amendments relating to other Byelaws as suggested by the tramway company.

The company was to be allowed, under Byelaw 8, to refuse to transport any person with an infectious disease on any of their services. Byelaw 11 was to be reworded to allow the company to refuse to sell return tickets, instead requiring that a single ticket be purchased by all passengers for each journey on the tramway.

The requirement to restrict boarding and leaving the carriages on the tramway was challenged by the tramway company due to the design of the cars that had been purchased or ordered for use on passenger services. Each car was divided in the centre, separating the compartment into first and second class saloons, although a gangway was provided to allow the conductor to pass to sell tickets and deal with queries and emergencies. The BoT sent a final letter, dated 1 November, in which allowance was made for the design of the cars in the required wording of Byelaw 2. The second byelaw was now to allow passengers to board or disembark from the tramcars only through the platform indicated by the conductor. The company was now able to finally print and advertise its conditions and byelaws at its stations.

With services about to start, the tramway was to be managed, in the short term, by a committee appointed by the directors during their 5 October meeting. James Clark, Thomas Tosland and Edward Davey had been appointed to run the tramway and were given management responsibility for decisions related to the daily working of the line. As with all the directors of the company at the time, their duties were undertaken simultaneously with them running their own businesses. James Clark was owner of the flour mill in Mill Street, Thomas Tosland ran a drapery business, whilst Edward Davey was a wine merchant. At the same time, the company hired its first two employees to work on the line. A driver was employed to attend to the horses working the line, and to drive the tramcar. The second new employee was a labourer who was employed to keep the tramway clear of obstructions and attend to any minor maintenance issues such as repairing fences and gates or cutting back undergrowth on the edges of the tramway.

The remedial works required by Colonel Hutchinson had been completed by the end of October, and he had been invited to return to Wantage to complete his supplementary inspection. Hutchinson wrote to Edward Ormond on 12 November to arrange for him to come to the tramway on the 16th of the month, arriving at Wantage Road station on the 1.06pm Down train. The inspection was completed, with Hutchinson finding that no further remedial works were required and he certified the tramway fit for public use in its entirety. The BoT wrote to the tramway confirming that conditions applied to the licence to run trains had been removed, and that the bond deposit lodged with the Court of Chancery had been released. The cheque refunding the deposit was received on 12 February 1876.

In late November, the collecting boxes being used by the conductors on the trams were viewed as being not fit for use with increasing volume of passengers paying fares aboard the tram and its creative view of where stops were. The company investigated how other busy tramways were dealing with high numbers of tickets, and it was decided to adopt the system in use on the tramways in Leeds. The directors, having discussed the system with their peers on the board of the Leeds tramway, purchased the requisite equipment and cash 'box' from the suppliers used by the tramway in Yorkshire.

As the year drew to a close, the tramway company invested in some additional improvements to the line. The second passenger tramcar was delivered to Wantage Road as promised by Starbuck Car and Wagon on 13 December 1875. The two cars were both horse drawn at the time, with the new slightly larger car requiring two horses, the smaller just one.

It was decided that, in view of the success of the passenger service, a better facility would be needed at the terminus in Wantage. A new building, referred to as 'The Station House' was to be erected, which would contain offices, a parcels office and waiting room for passengers. It was agreed by the company directors that the construction of the new building should proceed with invitations to tender for the work being circulated to local builders. Three tenders were received for the station house. John Wheeler and William Gregory, the builders of the existing facilities at Wantage, bid to construct the new building for the sum of £213 8s, with George and John Kent offering to build it for £194. The cheapest bid, which was duly selected by the tramway company, came from Whiting and Tosland who bid only £165 for the work. The contracts were signed on 2 December 1875, and construction began.

Chapter 5

GROWING PAINS

The new year of 1876 dawned with the Wantage Tramway Company encountering an unforeseen problem when James Sims, the goods manager, suddenly resigned. Sims demanded as compensation for the loss of his original business the entire takings for the goods traffic and half the profits on the parcels traffic for the first four months operation of the tramway. His plan was to restart his business without the GWR agency in competition with the tramway. The directors, in consideration of Sims having given up the agency with the GWR, agreed to meet his demands at their 18 January 1876 meeting. They informed Sims they regretted his decision to leave the tramway's employ in order to recommence his previous business.

February began equally poorly for the tramway when the trustees of the Besselsleigh Turnpike complained to the company that the rails had become higher than the level of the road in places. The trustees demanded that either the Wantage Tramway's rails were lowered at those points, not a simple request, or that the company paid the Turnpike Trust the sum of £75 for the level of the road to be raised to the height of the rails through an additional layer of macadam. The tramway company's managing committee decided to opt for the cheaper and quicker of the two options, paying the turnpike trustees to increase the road level. The payment was, however, only to be made once the road surface had been raised and inspected by a director of the tramway.

The company had a second difficulty to address following the departure of Sims from his post as goods manager at the end of January. The traffic was still growing, and the related workload was placing increasing demands upon the time of the management committee. The situation was discussed at a meeting of the directors on 5 February, at which the management committee was authorised to employ a replacement goods manager as soon as a suitable candidate could be found. The growth in freight traffic allowed the tramway to offer a concession to those companies who were placing a lot of freight with the company. A discount of 1 shilling, paid retrospectively, was to be given to any of the Wantage businesses who placed not less than 8 tons of goods with the company for carriage to Wantage Road, regardless of whether it was to be carried onwards by the GWR.

In March, the company found it necessary to arrange for the parcels office in Mill Street to be kept open throughout the day. The office had been closed for long

The Hughes Steam Tram, Wantage Tramway No. 4, near Wantage Road station with tramcars 1 and 3. The gradient on the Besselsleigh Road in the far distance is the bridge over the GWR's line, with the station building just visible and behind the second telegraph pole on the left. The driver's hand can just be seen on the regulator handle through the door which was often left open to improve the driver's sight line. The proximity of the rails to the road caused disagreements with the road agency at times. *John Scott-Morgan Collection*

periods during the day due to the management committee having to attend to other duties elsewhere on the tramway and in their own businesses. Even during the time that Sims had been goods manager, his offices in Market Place were often closed whilst he was out making deliveries and collections. It was resolved that the increasing demand for the tramway's parcel services, and the revenue generated justified the employment of a fulltime person to keep the office open throughout the day and to assist in booking in parcels and carrying out parcels to consignees in the town. Advertisements for the position were duly placed in the local newspapers. The position of goods manager was shortly filled, with George Stevenson's son, William, being appointed to the post on 27 March 1876 at an annual salary of £25. The very low salary suggests that William Stevenson's employment with the tramway may have been a part time role. Meanwhile the GWR and the Tramway had been clarifying many operational points which had arisen during their discussions.

The GWR and the Wantage Tramway Company signed a second contract which made the tramway delivery and collections agents for the railway as of 21 March 1876, in addition to the traffic agreement of 1875. The agency agreement specified that the tramway company as agents for the GWR must paint the words 'Great Western Railway, the Wantage Tramway Company Agents' clearly on the sides or fixed covers of all vehicles used in connection with delivery or collections. The tramway was prohibited from making any additional charges for deliveries or collections but would be paid an allowance proportional to the GWR's charges. The tramway was to collect the charges on behalf of the railway and deliver the monies to the stationmaster at Wantage Road station within seven days. All work was to done using preprinted stationery which would be supplied by the GWR.

The railway agreement, which was produced on a standard printed form so was clearly the usual arrangement between the GWR and its agents, specified who the tramway was to employ. A clause in the agency agreement stated that the tramway company must 'employ proper, civil, energetic and trustworthy men for the collection and delivery of parcels' with the addition that GWR officers 'may require the Agents to remove such Men, Horses, or Vehicles as they may consider to be objectionable or not sufficiently suitable'. The tramway directors agreed the content of the agency agreement with the chairman and Edward Ormond signing on behalf of the tramway, whilst James Grierson signed for the GWR.

The first six months after the tramway opened for traffic in October 1875 had produced some challenges to the company. The need for horses had caused delays in opening the tramway to passengers, the company finding it difficult to purchase suitable animals. The tramway company, in order to solve the problem, had made a similar agreement with Thomas Nunney to that already made with James Sims on 5 October. The company agreed a contract for Nunney to act as their agent for onward transport of passengers from Wantage Town station after he had ceased to run his public service into the town from Wantage Road station following the termination of his arrangement with the GWR. As a gesture of goodwill, which the Wantage Tramway Company entered into voluntarily, they would pay the sum of £100 to Nunney. Thomas Nunney was to have the sole right to operate public conveyances from Wantage Town passenger station and would be the only operator allowed to use the yard at the station for the operation of omnibuses. The tramway also gave Nunney passes for himself and all of his employees allowing them free travel on the tramcars.

In order to open the tramway, the company made an additional agreement with Nunney for him to continue to run his bus service for the first ten days. In return, Nunney would keep the entire takings for that time, costing the tramway company an estimated £20 with a profit of £6.

The tramway fares, which were often viewed as a little expensive by the people of Wantage and Grove, were set at 9d for first class, and 6d second class for single trips

A view into the tramway passenger terminal at Wantage with car 3 awaiting work. The structure of the overall roof added to the platform can clearly be seen. Without a run round loop, an arriving service would have to be backed out of the platform and run round in the yard, or another engine would need to take the cars out. Note the parcels barrow parked next to the wall and the gas lamps at the end and inside the train shed. *John Scott-Morgan Collection*

between Wantage Road and Wantage Town in either direction, with return fares at 1s 3d and 10d respectively. Travel between Wantage and Oxford Lane or Wantage Road and Grove Canal Bridge were 6d (10d return) first class or 4½d (8d return) in second class. Any travellers wishing to ride the tram between Wantage and Grove Canal Bridge, or Wantage Road and Oxford Lane would have to pay fares of 4d (6d return) in first, or 3d (4½d return) in second class. Goods charges had been set according to the rates allowed in the Provisional Order and Tramway Order which followed, which remained in force in the new year, however it was decided that the company needed to change its rates for the carriage of small parcels. The rates were to be reduced from 3d to 2d per package under 28lb weight, with packages weighing over 28lb but less than 56lb reducing from 4d to 3d each, and any heavier packages up to a maximum of 112lb to reduce in cost from 6d to 4d per parcel. The reduction had been possible as a result of the high number of parcels which were being carried on the tramway to and from Wantage Road station.

Goods traffic, with the exception of parcels which were carried in the passenger tramcars, had been transferred to the tramway with the road service ceasing, and were hauled over the line using draft horses purchased by the tramway company which were stabled in a purpose-built stable block in Wantage. The cart horses were only capable of hauling lighter loads, so the company had begun to enquire with locomotive builders and railway companies whether a small steam engine might be available to purchase for the goods traffic. It was envisaged that by the time a suitable locomotive had been found and purchased, its arrival on the tramway was likely to be after the BoT had authorised the use of steam on the line.

The early successes of the company were causing it some difficulties, particularly the demand for passenger services. The tramway directors stated in their second report to the shareholders, dated 1 April 1876, that only a single passenger car had been provided to cover the passenger services, which was proving insufficient. The passenger service timetable advertised at Wantage Town station aimed to meet certain GWR trains at Wantage Road. The tram aimed to arrive in time for the passengers to be able to purchase their tickets from the Great Western and catch the train they wanted. The tram would then be waiting for passengers to leave the train and board the tram, where they were able to purchase their ticket from the conductor in the tram car.

The service had proved impossible to deliver with only one horse drawn passenger tram car and with use of the Grantham Tram temporarily prohibited, the company had opted to purchase the omnibus which had previously been used by Thomas Nunney for £40. The tramway company's rail replacement bus service was viewed only as a temporary solution to the problem of offering the advertised service. The directors had ordered a second small passenger car immediately they had established that the service was undeliverable, but the Starbuck Car and Wagon Company would be unable to deliver it until 13 December 1875. The bus service would have to suffice until then, but as soon as the second car had been delivered in December, the bus service was to be suspended, with a two tram service delivering the timetabled service.

The operation of the tramway had, after the initial difficulties, run relatively smoothly. At the time of the directors' second report, the passenger cars were undertaking eight journeys in each direction between Wantage Town and Wantage Road stations. The company had begun to run an additional service on Wednesdays to accommodate the extra traders and visitors to the town market, the cars running nine return trips on that day. Punctuality was said to have generally been good, with no 'accidents or hindrances' and no other incidents causing any delays to passengers or goods. The tramway had received no complaints about the conduct of its staff either, which the directors were happy to report to the shareholders.

The company had increased their ability to run the service with the purchase of another six horses, four for passenger traffic and two draught horses to handle the goods traffic, although one horse had become sick and had been resting for a few weeks. By the time of the report, the horse was recovering well. The use of horses for the goods traffic restricted the company temporarily to the movement by rail of only light goods, the heavy traffic being sent by road using the company's own carts. The company was optimistic that the tramway would shortly receive permission from the BoT to use steam traction on the line according to the directors' report.

Whilst operationally the tramway appeared have started well, some issues had arisen with the rolling stock and infrastructure. The contractor's inability to complete the ballasting of the line before opening had meant that the 'roadway' between the rails during the winter had deteriorated but the contractor had consolidated the roadway and by March 1876 the entire tramway was once more in good order. The company had remained in good relations with the contractor throughout the resolution of the faults within the permanent way.

The company had experienced a problem with the passenger car quite early in its operational life. Once in use it had become clear that the new wheels, fitted to meet Colonel Hutchinson's recommendations, had been made by the foundry from a form of iron which was too soft for the conditions on the tramway. The regular use of the tramcars, coupled with the friction and grinding on the curves, particularly the sharp bends in the line leading onto the canal bridge, were causing the wheels to wear heavily, losing shape and becoming liable to derail. The company engineer had identified the problem and new sets of wheels to the correct profile and of a stronger material were due to be received in the late spring, shortly after the time of the second directors' report.

The work involved in meeting the requirements of Hutchinson's inspection regarding the wheels on the passenger tramcars and the Grantham Steam Tram and the improvements which followed after the tramway had opened was all undertaken by the Vale of the White Horse Iron Works company, by then renamed Gibbons and Gibbons. The two Gibbons after whom the company was named were Philip and Henry P. Gibbons, shareholders and directors of the tramway company.

The improvements to the tramcars wheels to allow them to run safely also involved the production of four new wheels and a set of brake gear for the Grantham Car, the rest of the work relating to the steam engine on board the tram and the changes to the cabs being undertaken by Nalder and Nalder and the Avonside Engine Company in Bristol. The new wheels for the Grantham Tram cost the tramway £13 17s. Tramcar No 1 received a new set of wheels costing £9 16s whilst tramcar No 2's wheels were found to only need turning down to produce the deeper flange needed at Wantage at a cost of £4 18s 6d to replace the shallow flange more suited to urban tramways with which the car had been built. The later adjustments to the tramcar wheels found after the tramway had opened was also undertaken by Gibbons and Gibbons, which cost a further £15. The company only presented their bill to the Wantage Tramway on 8 July 1876, once all the modifications had been completed.

Whilst completing the outstanding work to the infrastructure on their new tramway, the directors had taken the opportunity to use some of the revenue being earned to make improvements. A new waiting room had been built at Wantage Town, with the platform also being recovered at the same time. An additional siding had been laid in the yard at Wantage to accommodate the probable increase in freight traffic that was hoped to occur once steam traction had begun. The yard at Wantage had been prepared to make it possible to let out space for a coal yard to a local coal merchant which would earn rental revenue for the company.

The tramway had a somewhat creative view of what constituted a station from the very beginning. The only stations which officially existed on the tramway were at Wantage Town in Mill Street, and in the GWR yard at Wantage Road station. The company offered a part journey ticket, with the tram stopping at either side of Grove in each direction with the passengers alighting or boarding to or from the street. The two street stops were only ¾ of a mile apart but served passengers well. The trams also stopped where passengers had requested the driver or conductor to let them off the car, often outside their homes if they lived alongside the line. The tram would stop regularly for passengers who indicated that they wished to board the tram from the road side, private arrangements being made between locals and the tram crews. The directors were aware of the practice and supported it.

The Wantage Tramway Company was operating under an Act of Parliament and BoT licences which had been granted under the powers and regulations of the 1870 Tramways Act. The Tramways Act allowed the tram to stop at ad-hoc locations since it was incorporated as a 'street tramway' not a light railway. The regulations applying to City Corporation tramways thus applied to the Wantage Tramway and so the apparently random nature of the stopping points was perfectly normal. The frequent stops were no doubt convenient for those people who wanted to be able to use the tram from wherever they happened to be, but made the complete journey between Wantage and the GWR station somewhat slow.

Despite what can be viewed in retrospect as the teething troubles of a new tramway, and some challenging summer weather conditions, the company had achieved a good return over the first six months of the line's operation between 1 October 1875 and 31 March 1876. The winter of 1875-6 had been unusually wet and this, according to the tramway directors, had caused a general slackness of trade compared to expectations. The lower than expected traffic coupled with the need to undertake remedial works, and the purchase of an additional tramcar and

A local cartoon caricaturing the Wantage Tramway Company's passenger service. *Vale and Downland Museum*

horses had consumed a large part of the revenues earned by the tramway. The directors reported that, even so, they were extremely satisfied with the initial half year trading.

A few share subscribers had defaulted on calls for payment of the last amount called off against shares. Willaim Caudwell had paid only £15 of the £135 that was due against his share subscription, whilst L.W. Sansum (£100), Henry Cook (£50), Robert Sansum (£50), Alfred Kent (£20.10s), G. Kent (£45), George Stevenson (£90) and Edward Ormond (£90) had all failed to pay their contribution to the company finances which the directors had called upon. Most of the arrears were paid up soon after reminders had been sent out.

The company was still seeking to purchase steam locomotives for the tramway services as the summer progressed. George Stevenson was directed to contact three locomotive manufacturers, Hughes; Merryweather; and Shand & Mason, to enquire as to where their tramway locomotives could be seen for inspection. Stevenson was to report to the board of directors before accepting any terms offered by the manufacturers for hire and purchase of their engines. The company had also continued negotiations with Baxter regarding the use of the Grantham Patent Steam Tramcar throughout the spring and early summer, although the vehicle had already been delivered to Wantage and had been stored on a siding in the yard there since it had arrived the previous July.

The company and Baxter reached a mutually acceptable agreement for the Grantham tram's use on the line on 5 July. The Grantham tram was to be trialled on the tramway for three months, free of any charge from the owners, with a view to establishing if it was suitable to be adopted for the tramway's passenger traffic. Baxter had also been asked if he could approach the engineers who now owned the patent after Grantham's death, to determine if it would be possible for them to construct a similar vehicle for the tramway but with the engine located at one end, rather than in the middle as was the case with the existing steam tram. It seemed that the tramway had recognised the limitations of the existing design.

Stevenson had also been negotiating with Hughes of Loughborough and Merryweather and Sons with a view to bringing one of each of their locomotives to the tramway, Shand & Mason's tram locomotive having been discounted as unsuitable for use at Wantage. The Hughes engine was to be used on the goods traffic with occasional use on passenger trains when needed. Stevenson had been authorised to tender for the purchase of one of Hughes's locomotives, with the directors proposing paying £100 for the engine. Merryweather had offered to supply one of their tramway engines free of charge for one month, with the tramway paying the £5 cost of transporting it from London to Wantage Road. The engine could remain on trial at Wantage for a further two months for a charge of £1 per week, with Merryweather granting an option to purchase the locomotive for £500 at the end of the three month trial. Merryweather's offer was accepted.

Now that an agreement had finally been reached, the Grantham Tram could be tested. The tramway began by conducting limited trials in the yard at Wantage and the private part of the tramway in Wantage after the line had left the roadside. With the trials of the Grantham Tramcar giving the directors of the company great hope that the vehicle would be successful, it was to be introduced into revenue earning service as soon as possible, despite the tramway not having a licence to use steam on the line within the terms of their tramway order and BoT byelaws.

In the meantime, Edward Ormond and George Stevenson had met with Baxter and the Merryweather Engineering Company who informed the Wantage duo that a new steam tram could be constructed with the engine located at one end as requested. It would not be possible to sell the completed steam tram to the Wantage Tramway for less than £500. Ormond and Stevenson believed that the price suggested by the engineers was too expensive and informed them that they would not wish to proceed with engaging them to build a second steam tramcar for them. Baxter compromised and offered the company the right to purchase the existing Grantham Tram for £300. Upon their return to Wantage, it was agreed to offer Baxter £250 for it, subject to Colonel Hutchinson accepting its use on the tramway when he returned to undertake his 'steam' inspection for the BoT.

Baxter accepted the offer of £250 but rejected the clause relating to Hutchinson's inspection. The tramway, in need of steam traction, accepted Baxter's terms and agreed to purchase the Grantham car on 13 September 1876. The tramway company then planned to sell two of its horses which had been used for hauling the passenger tramcars once the purchase had been completed. It seems that the directors were anxious to get the Grantham Tram into service as quickly as possible.

The Grantham Car, despite claims which have been made to the contrary, had not been banned from operating on the tramway by company directors nervous about its use as already discussed. It had been in revenue earning service carrying passengers since the agreement for its trial had been agreed by the tramway. The BoT had turned a blind eye to its use despite the Wantage Tramway Order specifying that only animal power was to be used on the line. The BoT, in view of its own plans to seek a parliamentary amendment to the Tramway Act to permit the use of steam and other forms of mechanical locomotion by default, was keen to observe how the Wantage Tramway Company's experiment progressed.

Hughes Tram, Wantage Tramway No. 4, at Wantage, probably around 1920, with the skirts removed. It is noticeable that the connecting rods are still absent and the locomotive is running as a 2-2-0. Also noticeable is the long chimney and the toe hole in the front skirt in lieu of a step. *John Scott-Morgan Collection*

The trial and subsequent purchase of the Grantham Patent Steam Tramcar, with the imminent arrival of the Merryweather and Hughes engines, meant that the tramway needed an experienced driver to operate them. Advertisements were accordingly placed, with many applications being received with a selection of the men then being invited to Wantage for an interview. Thomas W. Hewitt was selected as the most impressive candidate, but before he was offered the position, the company took up his references. A favourable testimony was received from his primary referee, Edward Hailstone of Walton Hall in Wakefield, Yorkshire. Hailstone stated that Hewitt had worked for him for a long time, initially as a cowman, but when the estate had built their own railway, he had learned to operate and maintain a small steam engine. Hewitt was described as being respectable and honest, although a little slow and methodical in his work. The tramway company, as a result of the good references received, offered Hewitt the position, which he accepted. Hewitt relocated from Wakefield and the tramway had its first steam engine driver.

Before the company was officially able to begin running with steam, a second inspection would be required as indicated by Colonel Hutchinson in his first report during which a BoT inspector would reassess the tramway as built with a view to confirming its suitability for heavier steam locomotives or self-propelled tramcars. The inspection would especially need to ensure that all bridges were of sufficient strength to support the steam locomotives crossing them. The locomotives chosen for use on the tramway would require assessment on the same day before licences for their use on the Wantage Tramway would be granted. The BoT allocated the task once again to Colonel Hutchinson.

Hutchinson returned to the tramway in order to undertake the 'steam inspection' at the end of November, with the Grantham Steam Tramcar already in use on the tramway's passenger services. The inspection passed the line as fit for being worked by the steam tramcar, the bridge over the canal proving more than capable of handling the weight of the Grantham car which was run back and forth over it. Hutchinson wrote in his report that although in theory the tramway was suitable for steam locomotion, he wished to delay his final verdict until one of Merryweather's engines that were to be supplied at the request of the Wantage Tramway Company had arrived, at which point he would complete his inspection and report accordingly. When the first of the Merryweather engines had arrived, Hutchinson returned and completed his inspection of the line as well as the two steam engines now on the railway. The line was passed fit for steam locomotives, the bridge over the canal providing no cause for concern, but the results of the inspection on the locomotives themselves were not as welcome.

The steam tramcar was in use on both days that Hutchinson visited and he travelled on it between Wantage Road and Wantage Town and return, frequently requesting it to stop for him to perform inspections. It can only be imagined what the paying passengers thought of the even more frequent halts on the journey than usual.

The driver, wanting to make up for lost time, drove the Grantham Tram well over the speed limit of the tramway. The speeds approaching 20mph revealed to Hutchinson that the tram was not fitted with any form of speed governor, and the rigid wheelbase of the tramcar made the journey extremely uncomfortable for the passengers.

Hutchinson was very unforgiving of the Grantham Patent Steam Tramcar when he inspected it after his eventful journey on the tramway. His report was scathing in the extreme, finding it was completely unsuitable for passenger carrying services. Whilst Hutchinson stated that the Grantham car met its purpose as a passenger carrying vehicle, it had numerous faults which required urgent rectification to meet the BoT's safety requirements. The Grantham car had no self-acting speed regulator which shut off steam or applied the brakes if the car was travelling at excessive speed over the 10mph limit imposed on tramways at the time.

The Grantham Car was found to have no speed indicator, and Hutchinson raised a secondary concern that it had no means of recording the speeds at which the car had travelled. Whilst this requirement was a common byelaw imposed by the BoT, no such devices existed in 1876, so how the BoT expected tramways to meet their demand has been lost to posterity. Hutchinson was not satisfied with the arrangement of the machine at rail level, commenting in his report that no fender had been fitted at the ends of the car with which any obstructions could be pushed aside. Furthermore, no 'special bell' had been fitted to the steam tramcar by which the driver could sound a warning to other road users as the tram approached. The driver's cabs at each end of the vehicle were criticised heavily, Hutchinson unhappy that the driver's controls 'for operating the machinery' were not separated from the entrances to the car.

The Merryweather engine was less harshly treated, but Hutchinson still requested that some improvement should be made to the locomotive. The report stated that the engine, whilst having no self-acting speed regulator, was equipped with a governor on the exhauster which prevented the locomotive exceeding a speed of 11mph, which Hutchinson reported would be satisfactory. The locomotive was fitted with a speedometer, but Hutchinson observed that it seemed not to be working during tests undertaken for his inspection, although a simple repair undertaken by the company fitter ought to resolve the fault with no need for any further checks on his behalf. Hutchinson did have two small concerns which were raised in his report, although he believed that they were simple to resolve and again did not require him to reinspect the engine. The locomotive was fitted with front and rear fenders, but the shape needed improvement and the locomotive appeared to emit a large amount of steam which needed to be resolved. Hutchinson wrote that both these faults could be resolved through local maintenance and no further action was required on either his or the BoT's part. The same did not apply to the Grantham Tram.

Upon receipt of Hutchinson's inspection report, the BoT wrote to the tramway company. In their letter dated 2 December 1875, the BoT had stated that the Grantham

Steam Tramcar was not suitable for use on the Wantage Tramway and until such time as all of the faults found by Colonel Hutchinson had been resolved, the BoT was not prepared to sanction use of the Grantham Car. The directors of the tramway immediately responded to receipt of the BoT's letter by banning any further use of the Grantham Steam Tramcar on the Wantage Tramway. The content of the BoT's letter was what drove the directors to prohibit the Grantham Tram from running on the line, not reticence to use steam on the tramway as has been reported elsewhere.

The Grantham car has attracted something of a mythical status linking Loyd Lindsay with its creation, although this appears to be untrue. The steam tram was in fact a joint venture between the inventor and patent holder James Grantham and the Merryweather engineering company. The second decision by the directors to not allow the use of the Grantham Car on the tramway was purely due to the BoT inspector's refusal to licence it until remedial work had been undertaken upon it. There seems to be no evidence in the directors' meeting minute books and the company correspondence files to support the *North Wilts Herald* article of 18 September 1875, often been used to explain the delay in introducing the Grantham car into traffic. The directors seemed to have had no qualms about experimenting with new forms of traction, as other experiments were undertaken which will be described later in this book. It seems that the board were anxious to get steam traction onto the tramway as quickly and economically as possible from the outset, and the comments in the *Herald* seem not to be supported by the company archives.

The company still believed in the possibility offered by the Grantham Tram and arranged for a programme of private additional trials in the yard and line within the Tramway Company's land to determine any additional improvements that might be made in addition to those demanded in the BoT report. The directors were clearly not reluctant to use the Grantham Tram and their support for Grantham's invention can clearly be seen in the decision they made not to give up on the vehicle but to enlist the services of the local engineering firm Nalder and Nalder in Challow to inspect and improve the steam tramcar.

Thomas Nalder visited the tramway's depot at Wantage in October 1876 to inspect the Grantham Car and report on the work that the tramway company would have to undertake to resolve the faults found by Colonel Hutchinson. Nalder wrote to the directors on 10 October 1876 outlining his findings. Nalder's report stated that the car itself was just a fixed wheelbase ordinary tramway car which could be found in everyday use on tramways across Britain hauled by horses. The only difference was in the addition of a steam engine within the car to allow it to propel itself.

Nalder was generally impressed with the mechanical arrangements that had been made, especially the siting of the boiler and its engines in the centre of the car, spreading the weight across both fixed axles and all four travelling wheels. He was less impressed by the operation of the system. With the driver located at the controls in the front left corner of the car as it travelled, it was impossible to add fuel or water to the boiler without leaving his post whilst the tram was in motion or stopping the car on the line. His recommendation to the directors of the Wantage Tramway Company was that they should hire a stoker to attend to the care of the fireboxes and boilers on the Grantham Car, and thus release the driver to concentrate on his duties alone.

The boiler was viewed as being more than adequate for the demands that would be placed upon it, but the water supply on board was a little small, and a larger tank should be fitted. The system of refilling the boiler was however criticised by Nalder, who believed that the pump was too large for the piping and joints on the water system which it would be likely to burst. The pump was also subject to regular stoppage, with the only way to restart it being to dismantle part of the live steam feed into it. The pump was to be replaced with a smaller pump with a flywheel, both of which ought to be in sight of the driver. Nalder recommended that the replacement pump should be able to operated manually in an emergency, which was not possible on the existing water replenishment system.

The steam space in the boilers was viewed as too small, with Nalder observing water frequently being carried over into the cylinders (priming). The exhaust chamber for steam needed to be enlarged and arrangements made to condense the exhaust steam and return it to the water tank.

The footboard that the car had been built with was too small and extremely dangerous, Nalder recommending that it ought to be replaced urgently by an 18in wide board with a new handrail which extended the length of the car on the offside of the car to allow the driver to safely move from one end to the other.

The report was scathing of the brakes with which the tramcar had been equipped by the builders. Nalder found that the Grantham Car required 120 yards for the driver to bring it to a halt, with sixty passengers on board at line speed. The system of foot levers, which would have been more than sufficient for a horse drawn vehicle, were to be removed and be replaced with a handwheel and gearing as was more usual on railway locomotives at the time.

The last recommendation that the report made was for a signalling system to be installed on the tram to allow the conductor to attract the attention of the driver in an emergency, with the conductor also able to apply the brake himself and at the same time signal the driver to shut off steam. The Grantham Tramcar was supposed to be driven at around 10mph, but the tramway driver often drove it at between 15 and 20mph which was viewed as dangerous due to the combination of fixed wheel base and sharp curves on the tramway. The latter point was one which Colonel Hutchinson had also noted. Nalder's recommended solution to the speed issue was that the car should be equipped with an automatic brake running off a governor on one axle.

The Nalder and Hutchinson reports outline precisely why the Grantham Car was temporarily banned from use by the company directors following Colonel

Hutchinson's inspection and the BoT's reticence to licence its use on public services. The problems identified in detail by Thomas Nalder's report appear to have been caused by the experimental nature of the vehicle and its mechanical infancy rather than by any fundamental flaw in its design.

Whilst the improvements were being made to the Grantham Car, both the Merryweather engines were having improvements made to meet the recommendations in Hutchinson's report. The modified Grantham Steam Tramcar was returned to Wantage and Colonel Hutchinson reinspected it. The modifications were found to resolve all of his concerns and he reported accordingly to the BoT. The tramway was able to use the Grantham Car now that the faults with the prototype vehicle had been resolved and the tram was certified as being safe by Colonel Hutchinson. The tramcar was introduced into service and ran alongside the Merryweather engines before being replaced by a locomotive.

The fate of the Grantham car appears to have attracted some discussion, with some authors suggesting that the car was sold and ran on another tramway for a time. This is sadly untrue; the actual fate of the Grantham experimental tram will be discussed in a later chapter.

The company was finally granted a second provisional order in 1876 which amended the initial order, the Act of Parliament and the articles and association of the Wantage Tramway Company. A new clause was agreed between the Wantage Tramway Company and the BoT which allowed for the 'carriages to be moved by animal or steam power'. Importantly the clause also expanded upon the use of steam, allowing carriages to either be self-propelled by a steam engine located within the carriage or by a 'locomotive steam or other engines'. These locomotives were allowed to either haul or propel the carriages along the tramway, although they would always only haul the trains on the tramway in practice.

The BoT made it very clear to the Wantage Tramway Company that the amendment order was the first to allow any tramway to use any form of mechanical traction on their lines, and especially the first to allow the use of steam. As such, there would be a set of exclusions which would need to be added to the byelaws of the company before the tramway could officially use steam, although other forms of mechanical traction would also be allowed within the byelaws once passed. The allowances made for alternative forms of traction would allow the tramway company to experiment with some interesting traction machines later, but for now the focus was purely upon steam.

The Wantage Tramway already had one of Merryweather's tram locomotives on the line, as well as the disappointing Grantham Patent Steam Tramcar. The company had spoken to Merryweather and Sons after the inspection visit from Colonel Hutchinson with the intention of asking the London-based engineers to supply the second of their tram engines which had been tentatively agreed prior to the Colonel's visit. Merryweather had been informed of the outcome of the inspection and the minor problems which had been found with their locomotive at Wantage. In respect of the recommendations made by Hutchinson, the tramway asked Merryweather to arrange to add the necessary improvements to the second locomotive according to a specification which they had drawn up from the Colonel's report. The delivery of the locomotive was considerably delayed but not only by the addition of the improvements required to meet the BoT's requirements. Further delays were caused by Merryweather opting to make additional changes of their own to the locomotive. When the engineering company was prompted to deliver the locomotive to Wantage as soon as possible, Merryweather wrote back to Edward Ormond to explain why delivery was being further delayed. The engineers wrote on 7 November, explaining that the continuing delayed arrival of the locomotive was being caused by the addition to 'your engine' of a fresh improvement, although what this improvement was they did not say. They admitted they were causing the delay but were 'anxious to send you as perfect a machine as possible and have availed ourselves of every improvement which our experience has pointed out'. The locomotive was eventually delivered to Wantage, taking the tramway's No. 3 which had been allocated to it before the Hughes engine had arrived.

The company earnings, despite the pessimism of the directors in that regard, had in reality been quite good. The passenger traffic had earned the company £305 18s 8½d, with goods contributing £891 14s 8½d and an additional income from tolls and incidental charges amounting to £156 2s 9d and £47 18s respectively. The traffic charges payable to the GWR amounted to £789 of the goods revenue however, with salaries and wages amounting to £98 for the half year. The company had also spent £149 on stabling and feed for the horses, and a further £55 on new rails and taxing of vehicles to run on the roads. The company owed £150 for tolls collected on behalf of the Besselsleigh Turnpike Trust. The company was however profitable, and with £158 available for potential dividends the directors authorised a dividend of 2 per cent on the value of each share to be paid to the shareholders with the remaining 11s 2½d allocated to the company reserve account.

The finances of the company had received an injection of cash during the late 1870s, the November 1876 Annual List of Members and Summary of Capital and Shares of the tramway showed that 700 additional shares had been taken up, with 1634 shares now paid for bringing the total share capitalisation of the company to £8170. The majority of the additional capital had been invested by the now knighted Colonel Sir Robert Loyd Lindsay with 480 shares (£2,400), his wife Harriet with 80 shares (£400) which had been transferred from the ownership of Lord Overstone, her father, who had passed away in November 1883, and their son, Charles who had purchased 300 shares (£1,500). Edward Ormond had increased his shareholding in the tramway company, owning 50 shares (£250) by the time of the annual report.

The first few months of the operation of the tramway had been relatively successful despite having to resort to road haulage through the offices of Messrs. Sims and

Nunney. The first annual company accounts for the 1876 financial year presented to the shareholders on 1 November 1876 showed that the company had performed well and was well capitalised. The capital invested in the company through the purchase of 1,634 (£8,170) of the 2,000 shares offered, and Lord Overstone's loan (£1,800) had proved sufficient to support the construction and first year's operation of the new tramway.

The contractor Philip Ward's account had been settled in full, the contracted amount of £4,201 having been paid, with an additional sum of £565 5s 4d paid to cover the additional works needed during construction. The GWR had been paid for their work in laying the rails and completing the remaining works, including the junction with the Great Western's main line, within Wantage Road station yard, an invoice for £550 received from Paddington being settled in full. Edward Walmisley's account for his work in Westminster associated with gaining the initial tramway order had also been paid, a cheque for £286 being sent to him. The account for Walmisley's work during the extension of the order to include permissions to use steam traction of the Wantage Tramway had not yet been submitted by the year end but the company had been informed that the outstanding liability would amount to £140 19s 6d, but that would be paid in the next financial year, allowance being made in the accounts for the information of the shareholders. Finally, George Stevenson's account for his services had been settled, a cheque for £300 10s being presented to him in full settlement.

The company accounts show the investments which had been made in buildings and stock. The tramway had spent a total of £1907 10s on land, with a further £706 17s 9d spent on passenger tramcars, the horses needed to haul both the passenger cars and the goods traffic cost the company £292 11s. The tramway needed buildings which had been built by local contractors, the station house at Wantage Town and the adjoining passenger shed, platform and levelling off of the yard costing £184 and £327 15s 11d respectively, with George Stevenson submitting his 'engineer's charges' for both projects at an additional £90. A further £153 3s had been spent on repairing a house in Grove Street, adjoining the yard at Wantage which the company had purchased as part of the land which had been bought.

The last major expenditure before the tramway began operating involved the acquisition of Samuel Sims's carts and horses, as well as the purchase of the omnibus to complement the tram car services which had cost the company £116 11s. As a goodwill gesture to compensate Sims for the loss of his agency with the GWR, which the tramway was in no way obliged to pay, the company had made a payment of £100.

A few small purchases had been made once the tramway began to operate. It was found necessary to buy collecting boxes for the tram conductor for the collection of passenger fares on board the cars to assist the conductors in managing the cash which was being paid. The company also purchased sets of waterproof clothes for the tramcar drivers, who had no cabs on the horse drawn tramcars and were being exposed to all weathers. The next year of operation was looked forward to with much optimistic anticipation.

Chapter 6

THE EARLY DAYS OF THE WANTAGE TRAMWAY

Traffic on the tramway continued to grow in early 1877, passenger traffic steadily increased, while goods traffic grew quickly. The need to add another locomotive to the existing fleet was becoming increasingly urgent, although discussions with Hughes for the trial of one of their locomotives were progressing well.

The spring saw agreement reached with Hughes for one of their locomotives to be trialled at Wantage, and as a result the company entered into a contract with the local water company to provide a better supply to Wantage Town station yard.

The Hughes tram with passenger cars 1,2 and 3 at Wantage in the early 1890s. The first two cars were originally horse drawn, No. 1 operating the first services along the line. Car number 3 was built for steam haulage and is much bigger. Of note are the original bench seats on the roof of car 1, accessed via a vertical ladder which was put up to the open gap in the railings at either end of the car roof when someone wished to get up or down. It is notable that the connecting rods on the Hughes engine had already been removed as have the skirts. *John Scott-Morgan Collection*

The company completed the requisite works in April 1877 and all was then ready for the arrival of the engine. The Hughes tram engine duly arrived and was put to work.

The tramway company, always looking for an economic supply of reusable second hand materials, agreed to purchase all of the old Town Hall roof slates and floorboards which were being disposed of after the rebuilding and renovation of the building. The tramway company were planning on constructing a goods shed at Wantage Town during May and the recycled materials would be ideal, as well as being considerably cheaper than purchasing them new.

The tramway had carried goods for many regular customers on account, with the bill being paid at the end of each month. In the nineteenth century, these accounts were nearly always settled on time or at least within a short time after they were due, companies only occasionally having to deal with overdue debts. The threat of debtors' prison, and the social stigma associated with having been in that institution, meant that most accounts were paid in full quickly or if late often by arrangement with the company concerned. There were occasions when businesses and individuals found that they were unable to meet their liabilities. One such case affected the Wantage Tramway Company in April 1877.

The tramway had been carrying agricultural produce and machinery for Charles Edward Wells, a farmer. Wells's finances had become unmanageable by April with debts of £7,000 against assets of only £3,200, and as a result he had been unable to pay his account that had become due. Charles's brother Thomas wrote to the tramway company on 13 April, stating that he proposed to buy his brother's debts, including the money due to the tramway company, but only if all of the creditors accepted payment of ten shillings in the pound for the debts. The settlement would result in the company only receiving payment of half the outstanding amount but agreeing that receiving part was better than nothing, the tramway company accepted Thomas's proposal. The debt was settled shortly afterwards with the remaining half written off by the company.

The trials of the Hughes tram were not going as well as the Hughes company had hoped, although the tramway was more than satisfied with the performance of the engine. The company received a letter from Hughes which was discussed at the 2 May directors' meeting. Hughes stated that the engine had failed to carry out requirements, although the letter is unclear as to whether this refers to the locomotive being unable to perform the work to which it was being put, or whether the Wantage Tramway Company was not using it as often as they had agreed or Hughes had hoped. Regardless of the reasons for the perceived failure of the trial, Hughes had also stated in their letter that they would be removing the engine from the tramway at Wantage.

The tramway needed the locomotive, and the manager, committee and directors all believed that it was suitable for the company's needs. A letter was sent to Hughes accordingly on 17 May asking the company if they would be prepared to hire the locomotive to the tramway. Hughes declined, and the tramway wrote again offering to purchase the engine outright. A price of £620, plus the £2 5s hire charges for the trial, was agreed by the tramway company, on the condition that Hughes carry out some modifications to the locomotive. The chimney was to be 'heightened' to carry off steam and smoke clear of the passenger cars. A lever-controlled mechanism was to be added to allow the driver to divert the steam being exhausted from the cylinders with it to be turned back into a condensing stack, and then returned to the water tank. Hughes agreed and also lowered the price to £600. The Wantage Tramway took ownership of the locomotive which became Wantage Tramway Company No. 4, although the good feelings this generated on the board of directors would shortly be soured a little.

The company unexpectedly received a complaint and demand for delay repayment from one of their passengers. Thomas Cash of Aston in Birmingham wrote on 16 June 1877, complaining about the service on the tramway and demanding recompense. Cash was a travelling salesman who had visited the area to meet customers and otherwise conduct his business. He had used the tramway on 8 June and was a passenger from Wantage to Wantage Road where he claimed he had planned to catch the 5.58pm Down train. He said that the tram did not depart on time as 'the engine would not work' and so he and several of the other passengers on the tram missed their connection at Wantage Road. Whether the 'engine' was the Hughes or one of the Merryweather engines or whether he was referring to the 'engine' in the Grantham Steam Tram is understandably not mentioned but which was the guilty engine would later become clear in the correspondence that followed.

Nevertheless, the tram did run late and Cash claimed that having missed the evening train he had been unable to do the business he had intended nor had he been able to call where he had been advised to visit by his employers. He was therefore looking to the tramway for damages for his lost earnings and increased costs which he estimated at two guineas. He had left his card with the conductor of the tram car on which he had travelled as proof that he had been on the delayed service.

The tramway company responded with George Stevenson writing to Thoms Cash to deny any liability for Cash having missed his train at Wantage Road. Stevenson wrote that the tramway company apologised for the delay but could not be held responsible. The breakdown of the engine, Stevenson said, could not have been avoided. The engine in question was described as being a new locomotive and therefore could not be said to have been unfit to run. The reference suggests that the locomotive which failed at Wantage Town was the newly arrived second Merryweather tram engine No. 3.

Stevenson continued to refute Cash's claims, asking why he had not travelled on the Up train to Didcot which left Wantage shortly after the tram had arrived at

Wantage Road, and then taken the Down express which was scheduled to stop at his destination. Stevenson believed that, instead of waiting some time for the next Down train which stopped at Wantage Road, a detour via Didcot would have served Cash's purposes much better, allowing him to conduct his business that evening as planned. Stevenson informed Cash that he did not think that it would have cost more to travel to his destination via Didcot and the Down express. He would put it to the directors at their next meeting but doubted that they would entertain Cash's claims 'for one minute'.

Cash wrote to the tramway company again on 25 June stating that he adhered to his original claim. He wrote that despite Stevenson's thoughts, missing the train at Wantage Road had been 'a very serious matter' to him. Instructed by his employer to get home every week, the delay caused by missing the train had made it more awkward. Cash wrote that if he had been able to catch the 5.58pm Down train as planned, he would have been able to have done more business in Faringdon, and then caught the 8.45pm from there to Swindon. After staying overnight in Swindon, Cash claimed, he would have caught the early train from there back to Birmingham and been home early on Saturday morning.

Cash's letter went on to say that the delay to the tram caused him to miss his train, leaving him with a three hour wait at Wantage Road for the next Down train which upset his arrangements and so he went on to Oxford and then to his home. The disruption to his plans meant that, according to Cash, he had needed to return the following week to the places which he should have visited on the Friday. He also refuted Stevenson's claim that he could have achieved his aims by travelling up to Didcot on the next Up train and then back on the express. The next Up train did not leave Wantage Road, Cash wrote, until 6.39pm, arriving at Didcot at 6.58pm, whereas the Down fast left Didcot at 6.14pm.

Cash reiterated his claim for damages writing his arrangements had been upset purely by the tramway's 'car not performing the journey according to the time specified' in the company's timetable. Cash had taken legal advice and he said that he was intending to pursue his claim.

The company's costs had increased dramatically during the first half of 1877, with a considerable drain on the tramway's finances being felt by August. The great majority of the company's liabilities for future payment were related to the introduction of steam onto the line and improvements to accommodate increasing goods traffic. The work involved in the company's parliamentary agent Edward Walmisley submitting all of the necessary papers to the BoT and dealing with the correspondence related to the second provisional order for steam power on the tramway had cost £140 16s 6d. The cost of steam traction through purchase of the Hughes engine at £600 and the hire of the Merryweather engines for £71, had been added to by the cost of the remedial work required on the Grantham Tram.

Undertaken by the Avonside Engine Company in Bristol, the charge for the work to meet Colonel Hutchinson's specification amounted to a further £95, although not all of it would have to be paid. George Stevenson had also submitted a bill for his 'Engineer's Charges' amounting to £82 10s.

The purchase of the Grantham Patent Tramcar, Hughes Tram locomotive and passenger cars had consumed a large proportion of the company's available capital, and with only 1,634 of the shares offered in the company taken up, the tramway needed to find a way to inject some cash into the business in the short term. An extraordinary meeting of the shareholders was called for 5 September 1877 in the passenger waiting room at Wantage Town station on the tramway. A special resolution was passed allowing the directors to borrow up to £2,000 to cover the company's existing liabilities at a rate of interest not exceeding 4 per cent per year. The loans were to be taken out as a mortgage secured on the property of the company, adding to the existing mortgage loan of £1,800 at 4 per cent interest per year from Lord Overstone, which had previously been agreed by the shareholders. The second loan was duly taken out in two parts, with Gabriel Clarke and Edward Ormond. Clarke and Ormond loaned the company £1,000 at 4½ per cent interest, with Ormond loaning the tramway a further £400 at the same rate of interest. A further loan was later sourced from Edward Ormond on 22 May 1878 for an additional £400 also at 4½ per cent interest.

The entire authorised loan was not taken, only £1,400 was asked for and granted. The additional mortgage loan took the outstanding loan capitalisation of the company to a total of £3,000. The two loans were paid off through a special resolution at the 1885 shareholders' meeting. It was agreed that the loans would be paid back through the issue of 600 preference shares at a value of £5 each which were allocated to the people who originally provided the loan finance. The preference shares were set to give a guaranteed dividend of 4½ per cent of their value each year a dividend was available to be paid, giving these shareholders first priority on profits assigned to be used to pay dividends. Effectively, the payments covered the interest originally agreed on the loans, with the share capital representing the loan that the company would need to pay back at a future date. The share capital representing the loan was in effect only paid back when it was paid off to the holders of the preference shares at the time the company was wound up in 1946.

As the summer turned to winter, the volume of traffic being carried by the tramway was continuing to grow and the management committee could not devote enough time to the tramway alongside the demands of their own businesses. It very quickly become clear to everyone involved in the Wantage Tramway Company that a general manager was required to oversee the operation of the line on a daily basis. It was agreed to hire a new manager at the directors' meeting early in September, with advertisements for the position being placed in national engineering journals shortly afterwards.

One applicant wrote to the tramway offices requesting information. The correspondence gives an interesting perspective on what the tramway directors believed the manager's role was to be and how the staff recruitment process generally progressed in the nineteenth century. Thomas Barnes of Chorley in Lancashire wrote to Walter R.H. Barker, then the chairman of the tramway company. In the letter, dated 1 September 1877, Barnes described himself as a professional engineer who was 'very well acquainted' with the management of men as well as undertaking the routine duties of an office clerk. He requested that Barker send him a description of the expected duties of the tramway manager, and what salary the role would attract.

The chairman clearly responded, since Barnes sent another letter to him on 5 September. Thanking Barker for the response and information regarding the duties expected, Barnes went on to explain that he was conversant with them all. He stated that he was in possession of engineer's certificates and could provide further certificates from his recent employers referencing his 'integrity and ability'. Barnes commented in his letter that he thought that a wage 30 shillings a week (£78 per year) was very small for such a responsible job. Barker had apparently mentioned that the role attracted a share of the tramway's profits, and Barnes requested confirmation and guarantees of that commitment.

Barnes was not selected for interview, probably as a result of his demand for a higher wage. He wrote to Barker again on 22 September chasing for a response as he had not heard anything back from the Wantage Tramway Company about the job. He wrote that he believed he was well qualified for the job, with his experience justifying a higher rate of pay than the tramway had suggested. The tramway directors did not agree.

A large number of other applications were received, from which five were selected, all of whom were invited to come to Wantage for interview, with the company offering to pay for their second class rail travel. The men attended the 26 September directors' meeting in Wantage, during which they were interviewed by those present. The five men were H.C. Groz, a 31-year-old engineer from South Hackney; R. Waller Krug, a 29-year-old engineer from Brighton; Sergeant Major W. Pearson aged 41 and seemingly about to retire from service with the Royal Engineers; John G. Spencer, a 28-year-old engineer from Cardiff; and J.G. Scarlett, an engineer whose age and current place of residence were not mentioned in the meeting minutes. The directors viewed Spencer as the most suitable candidate, who accepted when offered the job.

The company employed Spencer as its first permanent general manager, appointing him at a salary of £75 per year. The 28-year-old had been living and working in Swansea where he owned his own home with his wife. He had been employed as an engineer in the town but was looking for a new opportunity 'with the prospect of advancement'. Interviewing well, the directors had offered Spencer the job of manager subject to receiving acceptable references. These were duly received from

Dr Taylor (Mayor of Swansea), his father John Spencer senior (Hotel Keeper) and the Reverend Thomas Walters of Llansamlet, a suburb of Swansea.

At the time Spencer was recruited, as was customary for many Victorian businesses, an insurance guarantee was taken out with the Guarantee Society, London, which had been empowered by a special Act of Parliament in 1842. The company took out the policy for a premium of £5 15s for twelve months with an extension for a second twelve months for a premium of £5. The policy covered the company from the day Spencer took up his position against any losses experienced by the tramway that were caused by 'fraud or dishonesty' on the part of the manager up to a value of £250. The policy limited the amount of cash that could be held by Spencer on behalf of the company to a maximum of £100 for at the most one week.

Spencer soon had two obstacles to deal with on the tramway. His first challenge arose when he received a legal summons for the tramway to attend court. The summons was passed on to Edward Ormond to deal with in his roles both as company secretary and the company's solicitor. The court summons had been sent from the County Court of Berkshire, requiring representatives of the Wantage Tramway Company to attend the next session of the court to be held in Wantage on 23 October 1877. The summons stated the case as 'Thomas Cash, Traveller for G.L Easton, Coal, Coke and Lime, Iron and Iron Ore Merchants of Trowbridge, Wiltshire. Vs Wantage Tramway Company Limited of Wantage'. The case to answer was for 'detention and loss of time owing to break down of steam tram car on June 8, 1877'. The claim was for damages of £2 2s plus costs of 4 shillings. The registrar of the court was Llewellyn Jotcham, a shareholder and future director of the tramway company. It seems that the company lost the case and had to pay Cash's damages.

The second problem occurred after an incident in which the Grantham Steam Tram had collided with the Hughes engine in late October causing minor damage to both and causing mild injuries to some of the passengers on board the Grantham car. The incident had been caused by the operating practices in force on the tramway. The Grantham tram had departed from Wantage Road after the GWR train had left the mainline station and headed for Wantage Town. At the same time, the Hughes engine had left Wantage Town heading for the GWR station. At a meeting on 5 November at which the incident was investigated, it was agreed by all present, including Spencer, that the Hughes engine ideally should not have left Wantage until the Grantham car had arrived.

Once departed from Wantage yard and knowing that he was likely to meet the Grantham car coming the other way, driver Redford should not have stopped the goods train on the line between the Wantage terminus and Grove Canal Bridge loop. The goods train should have been taken into the loop, only stopping for driver Redford to change the points to allow the train into the loop. The collision took place when the Grantham car driver suddenly found himself confronted by the halted

Hughes engine and its goods train, and despite applying the brake was unable to stop in time, hitting the stationary locomotive, although at low speed.

In order to prevent a recurrence of the incident, the regulations were changed. From 5 November 1877, no train or tram was to depart from either terminus without the direct permission of the manager John Spencer. In Spencer's absence, a nominated assistant would have the responsibility of acting as train despatcher, with a second responsible man being given the task at the other end of the line. The telephone line now installed between Wantage Town and Wantage Road must be used to ascertain the traffic on the line before any train was sent. The manager, or acting despatchers, were to keep a log book of all train movements, both arrivals and departures, from each terminus in the manner adopted by the GWR in their signal boxes. Passenger services were to be given priority, with goods engines following them down the line with their trains of wagons. Drivers of goods engines were instructed to keep 'at a proper distance' from the preceding passenger tramcars. If necessary to send conflicting movements, the tram or goods engine from Wantage Road was to go into the loop at Grove Canal Bridge and await the service from Wantage Town before continuing; this was only to be done when absolutely necessary.

Just after the crash, the company was approached by George Stevenson, who resigned as Superintendent of Repairs, a role which he had held since shortly after the tramway opened. The responsibility for maintenance of the line, its locomotives and rolling stock was placed directly under the control of the General Manager, to whom Stevenson had been reporting since Spencer had been appointed, from 5 November. Stevenson continued to serve the company as a consulting engineer after his resignation, however.

The tramway company still had to deal with the consequences of the Grantham car's collision with the Hughes engine. A letter was received by John Spencer from John Harris of Wantage who claimed that he had been thrown off the Grantham Steam Tram when the accident occurred. Harris said that he had been injured and claimed £10 in damages from the company. Spencer referred the correspondence to the directors, who agreed to pay Harris the sum of £5 in compensation for his experience and minor injuries, which Harris accepted.

The Hughes engine and Grantham Car were both soon repaired following the collision, but the company had realised that the Hughes engine might be more suited to the passenger services, with the Grantham Car being viewed as coming towards the end of its useful life on the tramway due to the limitation in the number of passengers it could carry. Its low power, which often caused it to struggle to cope with the incline into Wantage Town station, meant that it would be unable to haul a second passenger car. The more powerful Hughes locomotive was seen as being far more able to haul a train of passenger tramcars but that would mean it would not be available to haul the goods services.

The Hughes tram engine with a long train of three passenger cars travelling along the Besselsleigh Road sometime before 1900. The proximity of the line to the road is clear, and explains why occasionally the noise of an approaching tram may have caused a horse on the road to bolt. *Vale and Downland Museum. Wantage*

The directors of the Wantage Tramway Company considered their options for some time, with a decision being made at their 9 January 1878 meeting that, whilst the Grantham Car would be suitable in the short term, preparations should be made to provide passenger accommodation for the increasing number of people who were using the line through the provision of longer trains. The Hughes engine would eventually be moved over to hauling passenger services, with another engine to be purchased to replace it on goods services. The 'goods engine' would not be a tram, but rather the directors preferred to look to purchase a small 'ordinary steam locomotive' for the heavy goods traffic. It seems that the two Merryweather engines were no longer being used by the tramway and were not considered to be of any further use to the company by 1878. In view of the cessation of the Merryweather engines' use, the tramway company approached Merryweather for an adjustment to the agreed cost of hiring in the two engines, which the engineers agreed to reduce to £40, from the original £71 charge. The final hire fee was duly paid on 16 January 1878.

The search for a goods locomotive resulted in the company approaching the North Eastern Railway at York and the GWR at Swindon to enquire if either had a suitable engine that they would be willing to sell or lease to the tramway. The GWR responded that it did but would not lease it for less than £2 a day, preferring to sell the engine to the tramway. William Dean wrote to Edward Ormond on 15 February 1878 stating that he was prepared to recommend to the GWR directors that the engine in question be sold to the tramway company for £500. The engine had been built in 1871 by Avonside and Company of Bristol at a cost of £780 but had done very little work for the GWR. It seems the locomotive was too expensive for the tramway, who did not pursue the purchase.

The North Eastern did not have a suitable engine available but referred the approach to one of their locomotive suppliers, Black, Hawthorn and Company of Gateshead on Tyne. Black Hawthorn wrote to the tramway on 3 April saying that the North Eastern Railway Company had passed the tramway's approach on to them and that they did have a locomotive which might be suitable for which they enclosed a photograph, now sadly lost. The locomotive was a small, short wheelbase, four coupled tank engine, with 10in cylinders. It had been made by Black, Hawthorn and Company and was only five years old, and had been exchanged by the customer for whom it was built in return for a more powerful locomotive. At the time of the letter, Black Hawthorne said, it was undergoing a thorough overhaul and general repairs to return it to as new condition. The locomotive was described as being similar to engines supplied by Black Hawthorne to the North Eastern Railway and other main line railway companies. The locomotive was also described as having an 'awning' for the driver. Black Hawthorne offered to send a man with the engine to ensure it was delivered and shunted onto the tramway in good order. The price would be £550, with payment on delivery. It seems that the price was more than the tramway was willing to pay and the offer was declined.

The new year of 1878 saw the goods traffic on the tramway continue to increase dramatically. The parcels manager William Stevenson and his assistant were unable to cope with the increasing number of parcels being presented to the tramway company at the offices in Mill Street, the delivery of parcels which had arrived from Wantage Road and the increasing amount of heavy freight the company was carrying. In order to ensure the smooth running of the extremely lucrative goods and parcels traffic, something needed to be done.

The company approached James Sims once more asking him to return as 'goods manager and general clerk' to the tramway, overseeing the goods operation. His business must have been struggling to compete with the tramway services by then, since he agreed to do so. The contract was signed on 20 July 1878 and was to run for one year from 22 July with Sims being paid the sum of thirty shillings per week. On each six-month anniversary of the contract of employment, it would be automatically renewed for the period of a further six months on a newly negotiated wage unless either party had given notice in writing of their intention to end the agreement. Sims agreed and rejoined the company.

The search for a goods engine had been continuing with little success until the tramway company became aware of a potentially suitable locomotive which had become surplus to requirements as the works shunter at Crewe works on the London and North Western Railway (LNWR) after it had been replaced by a more modern engine. Purchase of the locomotive, LNWR No. 1863, was agreed between the two companies with it becoming the property of the tramway for a price of £350. The LNWR also agreed to undertake some 'special work' on the locomotive for an extra charge of £15 8s 1d. The locomotive was duly delivered to Wantage Road where it was allocated the tramway No. 5 and unofficially named 'Jane'.

As the year progressed, it was found that the canal bridge at Grove was beginning to exhibit structural problems. The tramway company approached the GWR for assistance, asking if they would be able to send one of their engineers to Wantage to inspect the bridge and recommend the remedial action that ought to be undertaken to repair and make safe the canal crossing. The GWR was unable to directly assist due to their own engineers' workloads, but the GWR Area Engineer at Swindon

A rural portrait of No. 5 on the line to the upper yard as it passed through the orchard. *John Scott-Morgan Collection*

wrote to the tramway on 2 August recommending an independent engineer, James Shopland, who lived in Old Swindon and would be able to undertake an inspection and make recommendations.

Shopland was duly engaged and found that two of the main girders which carried the wooden deck upon which the tramway sat were becoming stressed and showing early cracking. The two girders had to be replaced before the stress relieving cracks caused either or both of the girders to fail, potentially causing a train to fall into the canal. The tramway company engaged George D. Morton of Faringdon to provide and install the replacements. The two girders were to be supplied for the sum of £10 5s and installed under the supervision of the tramway company's engineer. The two girders described as being made up of two plates of 12in width and ½in depth with 3in angle iron rivetted on with ¾in rivets. It seems the preferred replacement girders were to be manufactured items rather than iron girders produced wholly within a foundry.

The company had purchased, as part of the land needed for the construction of the tramway terminus in Wantage, a house. The house had been in a poor state of repair and had been repaired on behalf of the company by a local builder and was now suitable to be leased to a new tenant. An agreement was reached on 28 September 1878 for William Church to occupy the house, but not the long shed and stables in the grounds of the house which the tramway reserved for their own use. The house in Grove Street was to be leased to Church at a rent of £20 per year, payable in four instalments in March, June, September and December with three months' notice required from either party to end the agreement.

The new year, 1879, began with the annual shareholders' meeting at Wantage on 23 January. A committee was appointed by the shareholders at the annual general meeting to look into possible improvements that might be made to the operation of the tramway now that it had been operating for over two years.

The shareholders' committee reported to the directors on 28 January and suggested two changes which both the committee and directors believed would be of benefit. The first change was to the fares and tolls then in place on the tramway. These were viewed as being unremunerative and as such should be increased. Goods rates were thus to be increased. Where freight charges of 3s per ton were being charged, they were to be increased to 3s 4d per ton; goods charged at 3s 6d increased to 4s 2d per ton; goods charged at 4s 6d to increase to 5s per ton. Goods charged not by weight but rather by the distance they would travel to their delivery station were also to have the charges increased. Coals were to be charged at an additional 2d per mile per ton, with timber loads to be increased by 6d per mile per ton with loads of iron, stone and bricks increasing by 3d per mile per ton.

The second change which the committee advocated involved the GWR. It was suggested that the directors should lobby the Great Western to induce them to discontinue the charges they made for 'hire' of the railway company's wagons whilst they were on the Wantage Tramway for onward transmission of the loads therein to customers in Wantage and for despatch of goods for onward carriage by the GWR. The directors agreed to attempt to convince the GWR to abandon the wagon charges laid down in the agency contract, although they reported that they were not overly optimistic that Paddington would agree. The GWR did not.

The early spring saw an interesting development. The board of directors instructed Edward Ormond to contact the GWR with a view to their potentially taking over the tramway. After an exchange to letters between the two companies, Ormond went to Swindon to meet with William Dean and, on a different day, with James Grierson at Paddington. The company wrote to the GWR on 13 March suggesting that the GWR take on the operation of the Wantage Tramway. A reply was received from Paddington, sent on 21 March.

The letter stated that the GWR had given serious consideration to the possibility of running the line, whilst Dean had reviewed the possible improvements that would be needed if the railway were to operate it under the powers of the various railway acts related to the GWR. Grierson's letter stated that the GWR board had concluded that at the 'present time' it would not be desirable for the railway to work the tramway. The letter continued that, whilst the GWR would no doubt be able to operate the tramway and maintain the stock and permanent way at less cost than the tramway were able to, the requirements contained in the byelaws which were not being carried out at the time would result in an increased cost, making it more expensive for the railway to run the services for the tramway. The conditions referred to most likely relate to the byelaws and conditions under which the GWR had to operate, which would have resulted in the need to maintain the permanent way to at least light railway standard, and probably would have involved the fencing off of the tramway from the road. The tramway thus remained independent.

Chapter 7

A VICTORIAN SUCCESS STORY

As 1880 began, the tramway embarked on a series of rail renewals under the instruction of George Stevenson. It had been found that some of the rails which had been used in the construction of the tramway had been defective, although the issue was a manufacturing fault rather than due to the contractor's work. The faulty stretches of rail were replaced with new, better quality lengths, which prevented any further minor derailments.

The expense of renewing the line added to the costs which had been concerning the directors during the previous two years. The account for the modifications which had been undertaken on the Grantham Steam Tram in 1877 had still not been paid in 1880, as the Bristol based company had experienced financial troubles and ceased trading. The liquidator of the company, Edwin Walker, had asked his solicitors, Fussell, Pritchard, Swann and Henderson, to contact the tramway with Edward Ormond receiving a letter, dated 28 May 1880, from them. The solicitors suggested that the tramway company need only settle a proportion of the outstanding debt, a payment of £50 in the next week, in favour of the Liquidators of the Avonside Engine Company sufficing to release the tramway from its obligations to the defunct company. The directors of the tramway agreed that this should be done, and a cheque was duly despatched to the solicitors. A receipt for the payment was received on 8 June with a letter confirming that the tramway company had been released from its obligations and the debt had been settled.

The tramway company embarked on a technological experiment during the summer of 1880. The company had been approached by the British representatives of the French Mékarski Pneumatic Engine Company. The Compressed Air Engines Company Limited had offered to supply two of their pneumatic locomotives with a view to trialling both of them on the tramway without charge for a period of three months. The directors had agreed to the trial and in accordance with advice from the Compressed Air Engines Company had informed the BoT in early July of 1880 of their intention to run the locomotives as an experiment and requested a licence to allow the locomotives to be used on the tram line. The BoT agreed, in a letter dated 16 July, for the experiment to take place, and informed the directors of the tramway that they would in due course despatch Colonel Hutchinson to inspect the new engines.

The provider of the pneumatic engines had in turn received permission from the tramway company to build, at their own cost, an engine house in the yard at Wantage Town station. The large new building was constructed to house both the air compressor and a stationary steam engine to power it. The building was completed with the engine and compressor installed and ready for use by the time of the BoT inspector's visit to the tramway to certify the arrangements and the locomotives.

Colonel Hutchinson arranged once more to come to Wantage, on this occasion to examine the pneumatic locomotives once they had arrived at Wantage in September. The inspector's report, dated 16 September 1880 stated that the two compressed air engines had already been numbered, as locomotives 1 and 2, despite these identifiers also being used by the Grantham Steam Tram and the first Merryweather engine. Hutchinson's report included a description of the locomotives. The four-wheel engines were powered by four compressed air cylinders, although only three were used in traffic, the fourth being a reserve for use in case of emergency to allow the locomotive to return to a recharging point. The compressed air motors had a design flaw which limited their use.

The locomotives could only run in one direction, which meant that they needed to be turned at each end of the journey and could not reverse at Wantage Town if they overshot the compressor house. A turntable was needed at both the termini of the tramway, which were provided and installed by the Compressed Air Engines Company at the same time they were installing the air compressor at Wantage.

The BoT finally agreed to the trial going ahead on 22 September 1880 and provided licences for the use of the two pneumatic engines, subject to the imposition of a set of conditions which had also been agreed as required by the tramway directors. Whilst working, the engines were not to travel above a speed of 8mph at any point on the line or in the yards at Wantage Town or Wantage Road stations. The locomotives were to be fitted with 'an efficient brake' and speedometer before being allowed to venture out onto the line. The BoT demanded that a governor should be fitted to the engines so that the brake was applied if the locomotives exceeded a speed of 12mph. The final condition to be applied to the working of the pneumatic engines was that all of the machinery, in line with other tramway locomotives, should be concealed behind protective plates above four inches from rail level. The Mékarski trials were not a success and the experiment ceased after the agreed three months.

The tramway continued to be successful during 1881. The company returned a healthy profit of £658 9s 8d from the company's activities during the financial year which ended on 1 October 1881. The shareholders again received a healthy return on their earlier investment in the tramway company. The sum available to pay dividends after a small part of the profits had been transferred to the reserve fund, was £645 8s 10d, meaning that the directors could approve another 5 per cent dividend to the shareholders. The tramway continued to be a good investment, although only 1,634 of the 2,000 shares available in the company had been purchased, the remaining 366 remaining unallocated throughout the life of the company. It seems that the profitability of the company and its annual returns was becoming a well-kept secret.

The new financial year began with another problem for the company to deal with when the tram collided with a carriage being driven along the Besselsleigh Road in early November, although damage to both carriage and locomotive were slight. The driver of the carriage claimed damages from the tramway for minor injuries which he received when he was thrown out onto the road but left the amount to be determined by the tramway directors. The tramway accepted the claim and paid the driver £3 11s 6d damages with a further £1 payment to cover his medical bills, both payments made on 10 November.

The tramway company became more open to the idea of issuing season tickets at a preferential rate during the 1880s. The first such arrangement was made in late November with William Reynolds who was residing near Wantage Road station. Reynolds and his family often used the tram to travel to Wantage for shopping and work, and he approached the directors of the Wantage Tramway Company to request a contract between the company and his family to cover their regular trips to and from the town for a fixed sum. The company directors agreed with the approach and offered the Reynolds family a three-month ticket for the sum of £2. The company applied one condition to the new season ticket arrangement, asking the Reynolds family not to travel when the tram car was full of other passengers who were paying the full fares. Reynolds agreed to the condition and the season ticket was issued, arrangements being made to inform the tramway's conductors and drivers of the new agreement.

The company also changed the working arrangements in the manager's office at Wantage Town at the same board meeting. The office was to remain manned at all times when the tramway was operating, with the 'boy' who had been employed as assistant to the manager to remain in the office as much as he could when the manager returned home for his meals. If anything occurred which needed the manager's urgent attention, the young man was to run to the manager's home to ask him to return to the office. The change in arrangements had been partly caused by the heavy winter rains which were causing the tramway to occasionally flood near William Jones's farm between Wantage and the Grove Canal Bridge.

The flooding was eventually resolved by constructing a new culvert under both the tramway and the Besselsleigh Road to take the water which was flowing off Jones's fields onto the tramway and road away from both transport links. The Highway Board at Wantage Urban District Council approved the work in early December with the tramway undertaking the construction of the new water culvert shortly afterwards. The new drainage solved the problem and both road and tramway remained passable in even the most inclement weather from then on.

The company was unfortunate in being involved in a second accident on the Besselsleigh Road in late December 1881. As a carriage driven, according to the board minutes by a 'Mr. Freddell', was crossing the tramway it was struck by the approaching tram. Freddell had attempted to beat the tram locomotive across the line despite it sounding the required warning, and unfortunately he didn't make the crossing in time. The carriage was damaged with Freddell receiving minor injuries, the horse was unharmed. The company, despite not being in any way liable for the accident offered to pay Freddell's claim for damages as a gesture of goodwill. Fredell received £2 5s to settle his claim.

Around the same time, Edward Ormond wrote to the Compressed Air Engine Company, inviting them to return to Wantage to 'level' the now empty buildings in the yard at Wantage Town which they had built to house the steam engine and compressors needed for the unsuccessful trial of the two pneumatic engines. The Compressed Air company duly wrote back agreeing to do so, or to pay the tramway company the sum of £5 if they would demolish the buildings on their behalf. The tramway agreed to demolish the buildings themselves and the buildings were cleared by the tramway company's men.

As the year ended, the tramway company received some unwelcome news from the GWR. Paddington informed the tramway that they were intending to reduce the charges made for the carriage of parcels in order to compete more effectively with other parcels carriers. The Great Western asked the Wantage Tramway Company to do likewise. The company responded by stating that whilst they were willing to help, they wanted to remind the GWR that the tramway company undertook the collection and delivery of parcels locally as well as carrying them along their line to Wantage Road and as such it would be difficult to reduce the tramway company's part of the carriage charges.

As the new year began, the company was involved in a serious incident in the yard at Wantage Town terminus. The goods engine, No. 5, was shunting trucks near the gasworks' gas house on Tuesday 3 January 1882 whilst James Palmer's maidservant was walking across the yard with a wheelbarrow full of clothes. The accident occurred when, according to the driver of No. 5 and other witnesses, the maidservant walking directly towards the locomotive which had whistled to warn her to move away from the moving engine, either would not or could not get out of the way of the locomotive which struck one corner of the wheelbarrow, fortunately missing the maidservant who escaped unharmed. The company authorised the manager to pay compensation to Palmer for the damage to his barrow, but also to remind him of the dangers inherent in his staff being in the yard and to ask him to remind them to take more care when about the tramway, and not to trespass in the yard as it was not to be used as a short cut.

In order to prevent a repeat of the incident, the company erected signs around the gas house, warning persons near the area that it was likely that a locomotive would be working the yard shunting trucks and that they needed to ensure it was safe to enter the area. The gas company was asked to move the wall near the gas house which restricted the clearances around the tramway's sidings away from the line so that three people could walk alongside each other between the wall and the rails. The gas company duly agreed to do so. Another notice was placed by the tramway at the entrance to the yard from Grove Street cautioning all persons against the dangers of trespassing on the tramway's line.

The directors and the manager had been looking to complement the Hughes tram with a new engine and had been in discussion with Matthews Engineers of Bristol to possibly take one of their tram locomotives on trial with a view to purchasing it if it proved successful on the line. An agreement was reached on 19 January, with Matthews offering to send one of its steam tram engines on loan for a month for trials on the company's passenger traffic. They also offered to 'lend a man' to manage and drive the locomotive whilst the tramway was trialling it. The directors informed Matthews that they were willing to agree to the arrangement on the understanding that the company may use the engine as often as they saw fit and that they could also discontinue the use of it at any time they wished during the trial period. Matthews agreed and arrangements were made for the locomotive to come to Wantage with its driver.

The tramway company's bad luck continued in late January, with another accident occurring on the line. James Dixon had been travelling along the road adjacent to the tramway on the evening of 20 January when the tram struck and damaged his gig, a light two wheel sprung carriage hauled by a single horse, usually with the driver's seat placed high above the level of the shafts for the horse. Dixon was catapulted out of the gig, which was damaged, but the horse remained unharmed. The tramway was presented with a claim for damages by Dixon, who asked for £3 15s compensation, which included repairs to his carriage. Witnesses, including the driver of the tram engine, stated that the gig appeared to be being driven recklessly, possibly attempting to race the tram. The tramway manager was authorised by the directors to make a small offer of recompense but not to accept any liability for the accident. Dixon received a letter from the tramway's manager, dated 26 January, in which he was informed that he should have better control of the gig, and that the tramway was in no way liable for the damage to the gig or to himself. The tramway did make some concession however, offering Dixon the sum of thirty shillings towards the repair of his carriage. Dixon accepted and the matter was closed.

The tramway continued to provide season tickets, although as in the case of the Reynold's family, only in response to a written application to the directors of the

The Matthews tram, No. 6, at Wantage. The track has been laid on longitudinal sleepers with iron tie bars preserving the gauge. Of interest is the point lever on the ground next to the locomotive. *Vale and Downland Museum, Wantage*

company. It was never possible to buy a season ticket for travel on the Wantage Tramway from either the conductors on the tram, the GWR ticket office at Wantage Road, or the tramway's offices in Mill Street in Wantage. In early February, Edward Davey approached the company asking them to grant him an annual ticket for unlimited travel between Wantage and Wantage Road stations. Davey was a coal merchant, who had taken up the lease of the coal yard in Wantage Road goods yard, as well as in the yard at Wantage Town. The company granted the season ticket to Davey for the sum of £5 for the year, also allowing Davey's coal agent, an employee, to travel for free when Davey was unable to attend to his coal trade at the GWR station personally. Davey agreed, paid the £5 and the pass was issued on 16 February. Davey, obviously travelling further afield for his business, shortly after being granted his annual pass approached William Gibbs, the GWR's station master at Wantage Road. Davey requested that in the case of the 5.45pm Up train being late, that the 6.31pm be detained to allow passengers to catch that service instead. Gibbs responded that he was willing to help the tramway as much as he could and

The slow speed of the tramway's passenger service, mainly due to the constant stops to pick up and drop off passengers at the roadside, meant that the service was often raced and also mocked. This cartoon relates to an incident which had become a local legend when Arthur Hitchcock, a chimney sweep with a wooden leg, is said to have raced the passenger service in his donkey cart – and won. *John Scott-Morgan Collection*

would detain the train if at all possible. Gibbs's actions show the good working relationship that the tramway company had with the GWR.

Davey seemed to have been not the only regular passenger using the trams. The increasing popularity of the tramway's passenger services allowed the company to reduce its fares for part journeys at the end of March. The fares between Wantage and Grove Bridge and Wantage Road to Oxford lane were reduced to 3d for first class and 2d for second class, whilst the fare for travel between Wantage and Oxford Lane or Wantage Road to Grove Bridge were reduced to 6d and 4d respectively.

The popularity of the passenger services led to local businesses approaching the tramway company to be allowed to place advertisements upon the passenger cars. The directors of the company, seeing an opportunity for an additional source of revenue, approved work to be undertaken to install advertising panels, referred to as 'tablets', inside each of the passenger vehicles. Each panel was to be used for advertising posters with a charge of 2s for each month that the advert remained on display from 13 April. The company also began to sell space on its tickets for advertising.

The tramway was beginning to experience regular minor derailments and rough riding on the line. The manager was instructed to look into the possibility of

An example of one of the tramway's passenger tickets. The tramway sold space on all of their tickets to local companies for advertising. *Vale and Downland Museum, Wantage*

replacing the now worn-out iron rails with steel equivalents. The tramway manager contacted the GWR's engineer, referred to in the correspondence only as Mr. Dunn, at Swindon for advice. Dunn recommended that the best replacement option for the tramway would be to use identical rails to those currently in use on the Swindon and Highworth Light Railway. The steel rails were made by Guest and Company at their works in Moorgate Street, London at a cost of £5 15s per ton. As an alternative for the line around and across the canal bridge, Dunn offered to supply good quality second hand 60lb per yard bridge rails for £5 per ton, delivered to Wantage Road station, which he advised would be an acceptable option for the tramway's use. The directors of the tramway authorised the purchase of one ton of rails from the GWR on 27 April.

No. 5 was in serious need of an overhaul by May which was becoming increasingly urgent. The company looked to hire in another locomotive to cover for No. 5 whilst it was undergoing repairs if one could be found at an agreeable rate. In the meantime the company had approached Swindon once more for help. The Great Western's chief mechanical engineer William Dean personally discussed the tramway company's requirements in a meeting in his office with Edward Ormond and James Spencer during early May 1882. Dean agreed that, despite the locomotive not being a GWR type, Swindon works would undertake to carry out all maintenance and renewals on the tramway's locomotive for a fee, which would advised once the locomotive had been inspected and the work needed had been identified.

The tramway approached other companies in an attempt to find a locomotive which they could hire in to cover goods services whilst No. 5 was away for repairs. The only other working locomotive on the tramway at the time was the Hughes tram, needed for the passenger service which meant that other arrangements needed to be made for the goods trains, even though the Matthews tram was due to arrive on the tramway in the near future but with it being restricted to the passenger services under the terms of the trial agreement it would not be able to undertake goods work even once it had arrived. The best offer that the tramway company received was from the Cambrian Iron Foundry in Wales who offered one of their locomotives for hire at the rate of 30s per day. The directors and manager thought this was too high and on 18 May decided that No. 5 would be sent to Swindon for repairs as soon as possible but no other engine would be hired in to replace it. The goods traffic was to be taken by road in the absence of the engine, even though this meant a large number of cart journeys would be required to replace each trip that the goods engine would have made.

The company, having purchased bridge rail from the GWR, approached Guest and Company asking them to tender for the supply of steel flange rails which would be used to replace the old iron rails on the tramway apart from in the area of the canal bridge. The steel foundry offered 50lb per yard steel flange rails and 68lb per yard steel bridge rails to the GWR's section for £5 10s per ton for both type of rail. The tramway agreed to order one ton of 50lb per yard steel flange rails for trial in late June 1882.

The company directors had agreed to award John Spencer a pay raise at their 18 May meeting. Spencer's wages were increased from £2 per week (£104 per year) to £2 5s per week (£117). The award was in recognition of his work over the previous year, adding to the rises he had already been receiving which had increased his salary from the £75 per year he had originally been employed on in 1877.

The tramway's goods engine had finally been sent to Swindon in mid-June. The news from Swindon after their inspection of No. 5 was not good. The cylinders of the locomotive had been found to be in very poor condition and extremely worn. Swindon advised the tramway company on 22 June that both the cylinders and steam chest were beyond repair and would need to be replaced. The GWR could provide new cylinders and a new steam chest, with William Dean writing that they would be happy to do so, but that there would be a cost. Dean suggested that the tramway company's manager come to Swindon to meet with him so they could discuss and agree how the work would be carried out.

Spencer attended Swindon works and met with Dean shortly afterwards. The two men agreed how the repairs to the tramway's engine should be undertaken. With Dean's designs for two new cylinders and a new steam chest being agreed, the GWR's Chief Locomotive Engineer gave instructions for work to begin on the repairs to No. 5. The extended absence of the locomotive from the tramway meant that the short-term expedient of using road haulage to carry the line's goods traffic was no longer suitable. It was decided that the Hughes engine, less capable of handling the heavy trains that No. 5 could manage with ease, would have to be used to haul shorter goods trains, limited to weights which it was capable of handling, between each of the day's busier passenger services. Additional trips would be scheduled where necessary at the discretion of John Spencer, in accordance with the freight waiting at either end of the tramway after each trip that the Hughes engine undertook with its wagons. Whilst the Hughes engine was engaged on the goods services, the company still had the Grantham Patent tram in service, which could be used on the more lightly used passenger services during the day, when its limited carrying capacity might be less of a problem.

The company sold its third season ticket on 22 June 1882. Edward Ormond was granted a season pass for both himself and all of his household at a cost of £1 per month or 5s per week if a shorter pass was desirable. The pass allowed any member of the Ormond household to travel on the line at any time.

The Wantage Tramway was still operating its freight services on account for its regular users. The company had learned from the situation which manifested around Charles Wells's debt in 1877, and the account books were regularly checked for overdue debts. On 31 August 1882, it was found that the carriage charges owed by James Hiskins had become overdue. The manager, in accordance with his instructions from the board, referred the debt to the company's directors who instructed him to approach Hiskins to settle it. Hiskins was asked by Spencer to pay the outstanding amount within two weeks, warning him that if payment were not received by the end of that time legal proceedings would be undertaken.

August saw the Matthews engine arrive at Wantage and it became the tramway's No. 6. The company directors instructed that it should immediately begin trials on the passenger services. The experiments were to be undertaken under the direct supervision of the tramway's manager. Spencer was to ride on the footplate and in the passenger cars to observe the performance of the engine and would report back to the directors on a regular basis. The arrival of the Matthews engine could not have occurred at a better time for the tramway. With No. 5 still away at Swindon, the new engine could replace the Hughes tram which was then available to work the goods trains along the line. The tramway's difficulties were thus solved for the short term until No. 5 returned, even though the Hughes tram was only capable of hauling lighter trains.

The approaches to the canal bridge at Grove were still causing trouble in 1882, despite earlier attempts to ease the approach curves to the bridge. In September it

The Matthews Tram engine No. 6, and passenger car No. 3 at Wantage. The bar coupling can be seen on the front of the engine, folded away. Of note is the somewhat uneven lettering on the side of the locomotive, the sign writer must have had a bad day. *John Scott-Morgan Collection*

was decided to improve the line at the bridge so that trams did not have to slow to walking pace to avoid being derailed on the sharp curves at either end of the bridge. The directors inspected the bridge with Philip Gibbons, suggesting that if the curve at the south side of the bridge were to be re-laid at a more generous radius, the tramway would be able to remove the existing speed restriction. The work was duly authorised and undertaken. Whilst the company was looking to improve its infrastructure, a tender was put out for the lowering of the road into the tramway yard in Wantage. A bid to do the work for £7 was accepted and contracts exchanged on 21 September.

A series of minor derailments had occurred on the tramway, particularly when the trams were passing through facing points in yards and on the passing loops on the line. In response to the reports being received from the tramway of these minor and easily rectified mishaps by the BoT, the board issued a new byelaw order. The order enforced that from 24 October, all trams were to slow to 4mph when approaching facing points, with the manager to inform all drivers of the new requirement.

The winter weather was particularly cold and wet from early November, and as a result the drivers requested waterproof greatcoats from the tramway which they

The Hughes Steam Tram, Wantage Tramway No. 4, at Grove canal bridge with two passenger cars. The road bridge can be seen behind the tram conductor who is standing on the road. The rails visible behind the train are the passing loop. The location is one of the official stops on the tramway, although the tram would regularly stop at random places along the line to pick up and let down passengers. *John Scott-Morgan Collection*

could wear when out of the cabs of their locomotives. The directors discussed the request at their 16 November meeting but delayed purchasing greatcoats for the drivers and the conductors until the manager had acquired a suitable sample. At the same meeting it was agreed that rather than charge William Weedon for repairs that the tramway company's workmen had made to the road leading into his coal wharf in Wantage, the company would accept a truck load of apples from his orchard. The manager was instructed to sell the fruit to passengers and the non-travelling public at the tramway stations to recoup the cost of repairs.

The tramway was pleased to see the goods engine return to the line. Swindon had completed the needed repairs and installed the new cylinders and steam chest so that when No. 5 returned to the line it was almost a new engine. The company now had three locomotives to cover its daily needs. No. 5 and the Hughes tram could cover the goods services, whilst the Matthews engine, still on the line, could share passenger trains with the Hughes locomotive.

The tramway company began the new year with an inspection of the outstanding debts due for carriage. The manager, Joseph Aldworth, who had replaced J.G. Spencer in 1880 following a period when he stood in as interim manager after Spencer had suddenly decided to return to his home in Cardiff, was instructed to approach all of those whose accounts were overdue. There were many individuals and companies who owed the company payment for the previous financial year's carriage, the debts having been outstanding since the end of September the previous year. Aldworth had been instructed to have all of the outstanding monies paid, so that the cash needed to pay the previous financial year's dividends was available in the company's bank account. Thought was given at the directors' meeting of 11 January 1883, as to how the company could in future ensure that carriage charges were paid promptly by all of its customers so that time and effort did not need to be spent chasing up overdue debts.

James Wilshire, the company's managing engineer who had recently taken over from George Stevenson in the role, had asked to speak to the board and was invited to do so at the same meeting. It seems that a disagreement had occurred between Aldworth and Wilshire. Wilshire asked the board whether they were satisfied with his management of the tramway's locomotives, rolling stock and infrastructure. He also asked the directors, in view of his disagreement with Aldworth, to clarify in writing both his and the general manager's duties. The directors assured Wilshire that they were more than happy with his work and agreed to define clearly the duties and authorities of the managing engineer and general manager.

A special meeting on 25 January was arranged with Wilshire, Aldworth and Edward Ormond in attendance. The two men were presented with the written definitions of their roles by Ormond to resolve the misunderstanding which had resulted in the disagreement between Wilshire and Aldworth. The General Manager had sole management of the men on the tramway, including the engine drivers whilst they were managing their locomotives in the yard at Wantage. The latter was in accordance with the regulations of the tramway put in place to avoid head-on collisions on the line. The General Manager was finally responsible for regulating traffic and reporting to the directors any obstruction on the part of the men to the smooth running of the tramway.

The Managing Engineer's duties were to have sole control of the repairs to engines and passenger cars as well as the maintenance of the line. The engineer had management oversight of the men whilst out working on the line. The engineer's assistant was to be overseen by the engineer until 7pm each day, when the General Manager may require him to deliver the passenger luggage which had been left with the company after the passengers had completed their journey to Wantage. The engineer was to be solely responsible for ensuring that the engines and cars needed for each day's work were made ready 'for the proper working of the traffic' as required by the general manager. The definition of each man's duties and responsibilities resolved the disagreement, both men working together with no further disputes from then on.

Around the same time as the difference of opinion between the tramway's two senior managers was being resolved, the drivers were finally provided with waterproof clothing for use in inclement weather. One greatcoat was provided for the driver of the passenger tram locomotive in very severe weather. The coat was to be kept on the engine during the winter, but during the summer was to be stored safely in the offices in the station house at Wantage. The driver was not to take the coat home at any time. There is no mention in the company correspondence files as to whether the goods driver or the conductors received greatcoats, but it seems likely that they were also provided with them.

The company was once again involved in an accident in February 1883. A truck belonging to the coal company providing fuel to the gas works was damaged when it was involved in a collision with John Pinnell's donkey cart whilst being shunted near the gas works. Pinnell's cart was crossing the yard and hit the moving wagon, damaging both vehicles, although driver and donkey remained unharmed. The fault for the accident was deemed to have been with Pinnell for not keeping a good look out whilst in the yard collecting his goods. No liability was accepted by the tramway, and as a result no compensation was required. The owner of the railway wagon opted to repair the light damage at the truck's next overhaul, with no charge being made to either Pinnell or the tramway.

The inclement weather had continued into February, with heavy rain causing the drains at the side of the tramway to fill and spill over onto the rails and the roadway. The local highway agency complained to the tramway manager that the drains by the side of the tramway were also blocked and suggested that the committee of the highways board should meet with the directors of the Wantage Tramway Company at the location of the problem to jointly inspect the drainage. It was agreed that something needed to be done to improve drainage, the original drains being unsuitable. The tramway purchased iron pipes and installed them under the tramway so that water could flow out of the drains and into the ditch at the side of the tram line from where it could flow away.

The GWR and the tramway company had been in long discussions about the work being undertaken in Wantage Road yard by the Wantage Tramway's goods engines. The tramway had asked the Great Western for their help in reducing the huge amount of shunting being undertaken by their engine, which included marshalling goods trains ready to be added to the GWR's pick up freight services to Didcot. They also claimed that sorting in-bound goods wagons was taking up a lot of the goods engine's time as well. The GWR responded in a letter dated 22 February. The GWR's goods manager informed the tramway company that the cause of the excessive shunting being undertaken by the tramway was the length of the siding provided by the tramway. As traffic had increased dramatically, the short siding which could only hold four trucks was insufficient and that thought should be given to the tramway engaging the railway to extend the length of the siding to allow longer trains of wagons to be placed onto the exchange siding, thus removing the need for the tramway engine to be continually shunting trucks around the yard and Wantage Road.

The tramway had other things to deal with in February apart from the constraints at Wantage Road. The goods engine had developed a fault which needed minor repairs that could be undertaken by the engineering manager. Whilst Wilshire was attending to No. 5, the Matthews tram engine broke some of its springs which would need replacing before it could return to work. The tramway was unexpectedly finding itself short of motive power, with only the Grantham Tram, with its limited passenger capacity, and Hughes tram available to cover both passenger and goods services as well as shunting the yards. Fortunately, the repairs to No. 5 did not take long, and Wilshire was able to put it back into service in a few days, the repairs to the Matthews tram took a little longer than expected, but the tramway found its motive power crisis had mainly been averted. For a few weeks, many people on the tramway were hoping that the Hughes engine would keep working without developing any faults of its own, which fortunately it did, not once failing to be available for traffic. The Hughes engine and No. 5 carried on faultlessly whilst the Matthews engine was out of traffic. The Grantham Steam Tram was still being used, but its inability to haul an additional passenger car, and limited capacity meant that it could not be used when the services were liable to be busier, but at least it was available for traffic.

The highway agency was back in touch with the tramway's directors again in late March. They now demanded that the drain gratings between the tramway line and the road should be cleaned and kept clear by the tramway company. They also informed the company that the road and tramway had been inspected, and the agency now required the tramway to lower the height of its rails to be at the same level of the road. The highways committee only offered one compromise in their 22 March letter; they would undertake to keep the ditch alongside the tramway clear but all other maintenance expenses were to be the responsibility of the tram company.

April arrived and with it the directors' spring meeting. The communication from the highways committee had caused the company's management to think about other improvements that could be made to the line whilst the weather was good. In order to achieve the needed repairs, and to meet the highways agency's demands, the company began to buy in track materials ready to relay the parts of the line which needed work. At their 5 April meeting, the directors also discussed the wages that were being paid to the 'Parcels Boy', S. Painter, and finding that his hard work over the last few months deserved a better reward, the directors agreed to increase the young man's wages from five to six shillings per week.

The tramway had arranged to purchase a redundant set of points and guard rails from the GWR for £15, as well as a load of wooden sleepers, to be used in adjusting the level of the line that the highways committee had complained about. The tramways other plans were not entirely proceeding to schedule. The Matthews tram had still not returned to service by 12 April when the board were informed that it was in a worse state of repair than expected. Wilshire had removed the wheel sets to replace the springs and whilst doing so had found that the engine needed all four wheels replacing. The repairs would therefore take considerably longer and needed the authorisation of the directors for Wilshire to order in the new wheels from Matthews. The tramway would not have a spare engine for a while longer, despite authority being given for Wilshire to undertake the necessary work on the engine at Wantage. The directors instructed Wilshire to keep a record of the hours involved in making the repairs, both by himself and his assistant fitter. The board intended to bill the cost back to Matthews, given that the engineers still owned the tram engine, and maintenance of the locomotive by the tramway company had not been agreed in the arrangements made to extend the trial of the Matthews machine. The lack of motive power may have partly contributed to a problem that the tramway seemed to be having during the late spring.

The company seemed to be having trouble with collection of goods during the early part of 1883. H. Sumerton of the GWR's goods managers office wrote to the tramway in April asking why the collection of goods had become irregular. A few of the consignors of freight had complained directly to the Great Western that their goods had not been collected quickly. The reason that the GWR had received the complaints was due to the paperwork stating that the tramway were agents for the railway, rather than carrying the freight in their own right, many people assuming that the GWR was operating the service, and thus complaining to the railway rather than the tramway company.

The GWR was concerned about the problem since it damaged their reputation. The railway's officer also raised a secondary concern which had arisen over the previous few months, also related to the goods traffic. It appears from the correspondence that the tramway was not obtaining signed consignment notes which proved that delivery of the related goods had been completed, and that the delivered goods were undamaged. The company was also not submitting the completed consignment notes to the GWR at Wantage Road as required in the agency agreement.

The management of the tramway seemed to be unaware of the lack of paperwork. The general manager was summoned to the next directors' meeting, which took place on 19 April. Aldworth reported that he was unaware that there had been a problem with the goods service, which as far as he had been aware had been functioning as normal. Aldworth was instructed to resolve the issue quickly and to respond to Sumerton's letter. The general manager promptly investigated the situation and found that George Ryman, the company's goods agent, had not been collecting signatures on delivery in order to complete his rounds more quickly. He had also not been returning to collect goods later in the day, preferring to perform collections as he was dropping off deliveries. If the shipment was not on his planned delivery route, he was not diverting to make the collection, instead leaving the goods uncollected until he was passing nearby on a delivery run another day.

Aldworth reprimanded the goods agent and instructed him that all booked collections must be brought to the tramway every day, and every delivery must be accompanied by a signature from a representative of the consignee without fail. Ryman was directed to ensure that he complied at all times with the GWR's regulations, and was reminded by James Sims, the tramway's goods manager, about what those were. Aldworth would, for a time, inspect the daily manifests to ensure that the goods traffic was being handled as the tramway expected. The general manager then proceeded to write to Sumerton to reassure him that the problem had been dealt with, and that no further delays would occur with the freight traffic unless unavoidable or caused by the consignor of the shipment. The issue did not recur.

The goods traffic on the tramway received a boost in June when the Post Office approached the company. In a letter dated 21 June, the Post Office asked the tramway company whether they would be prepared to carry the mails from Wantage Road into Wantage for delivery to the local postmaster, and if so, at what cost. The tramway manager was instructed to respond positively to the request, the directors being keen to take on the extra work and the revenue it would bring. Aldworth wrote back to the post office informing them that they would be prepared to collect the mails from the GWR at Wantage Road and carry them on to Wantage. The cost for providing the service would be 25s a week.

The Hughes tram engine with passenger car No. 3 at Wantage Road. The two gentleman standing at ground level with the trolley are about to load the parcels onto the car for carriage to Wantage Town. Just visible on the near side of the end balcony are mail sacks destined for the post office at Wantage. *John Scott-Morgan Collection*

The improvements to the line which had been discussed in April began in the summer as planned. A load of used sleepers had arrived at Wantage Road by 5 July from the GWR's Reading District permanent way depot, as had the rails which had been purchased from Swindon and Guest and Company in London. The managing engineer began work on various parts of the line, which could be quickly completed without stopping the tram service. He also had a major project which would have to be undertaken on Sundays when the tramway was not running. The local highways engineer had to be consulted about the intended works, as the easing of the curve required the footway on the offside of the Besselsleigh Road to be altered. The highways agency gave permission for the tramway to move the footway whilst the line was being re-laid. The easing of the curve on the approach to the canal bridge at Grove was completed over one weekend, the trams then being able to run at line speed through the bridge.

The permanent way complications had been resolved, but a new problem presented itself to the tramway. The Matthews tram had returned to work following the completion of the work to replace its springs and wheels but, almost immediately, the Hughes tram engine had failed. The engine had been worked hard whilst both the Matthews tram had been out of service and the goods engine, No. 5, had been away at Swindon for repairs. Once No. 5 returned to Wantage, the Hughes engine was able to return to just working the passenger services, but still had to work every day without a spare engine to allow it to be taken out of service for boiler washouts and minor maintenance. The hard work and age of the engine had finally caught up with the tramway in July 1883, when the locomotive was failed with a leaking firebox and other mechanical faults which needed rectification. Wilshire informed the general manager that the locomotive needed a new firebox, the current one being beyond repair due to serious cracking around the stays. Aldworth was told that the locomotive needed a lot of mechanical repairs that could be undertaken while the boiler was away for the repairs, but No. 4 would be out of traffic for some time. The Grantham tram would have to share passenger duties with No. 6 until the Hughes engine had been repaired, which was not ideal. The directors were duly informed that the motive power issue had occurred again.

The locomotive had been inspected by George Stevenson and again by Wilshire, both men agreeing that the locomotive was safe to use only on light work. The tramway sent the requirements away for estimates from local engineers and Hughes in Loughborough, as well as the GWR at Swindon. In the meantime, while the locomotive was waiting for its new firebox, a decision was made to continue to use the Hughes tram on goods services when needed, but an instruction was given to both the general manager and engineer that the locomotive should only be allowed to haul two full wagons from Wantage Road on each trip. No. 5 was to undertake as much of the freight work as possible, while the Matthews tram engine would handle the majority of the passenger services. The Grantham tram was to be used when a second service was timetabled to follow soon after a previous departure, when the Matthews engine did not have enough time to complete a return journey prior to the second service having to leave either terminus. Passengers would be encouraged to use the longer service hauled by the Matthews engine, unless they had urgent need to get to the other end of the line, when they could travel on the Grantham tram which would depart first. The changes to the passenger services would have to suffice in the short term until the Hughes engine had been repaired and could rejoin the fleet. No suitable locomotive was available at Swindon, so the company would have to manage as best it could without the Hughes engine.

An example of how well the tramway and GWR worked together at Wantage came about in August 1883. A complaint was received by the tramway in relation to a delivery that had been made in August. A Mr. Leveson complained that a consignment of manufactured perambulators which had been sent from Wantage had

The Matthews steam tram engine in the passenger shed at Wantage. The panel on the side of the locomotive under the word 'Tramway' is the skirt over the motion which was often left raised to give quick access for oiling up. The passenger is sitting on the end of tramcar No. 3. *John Scott-Morgan Collection*

arrived at their destination in London damaged. The consignment had been worth £9 9s, which his customer was not now prepared to pay, and so Leveson was seeking recompense for the whole value of his goods. The tramway could find no evidence that the damage had happened whilst the shipment was in their care. It appeared to have been handed over to the GWR in good condition, although it was difficult to ascertain if any damage had occurred early in the journey as no one at Wantage Road could recall the state the packages had been in when they arrived there. The GWR agreed that the damage could have happened anywhere but had definitely happened during transit. The two companies agreed that it would be fair to pay Leveson for the goods which had arrived damaged at Paddington, each company paying half the value. The claim was settled on 30 August with the tramway paying £4 14s 6d, and passing on a cheque from the GWR for the same amount.

The Hughes engine returned to full traffic towards the end of the year with a new firebox. Its return was very welcome. The freight traffic on the tramway had continued to expand, particularly coal traffic for the gas works and Wantage's households and industries. Coal traffic had mainly been provided by Edward Davey's coal merchant business. The coal traffic had been so regular that the directors of the tramway company, on 1 November, offered Davey a rebate of 2d per ton on the coal which had been carried on his behalf during the entirety of the previous financial year. The directors only made the offer to Davey, instructions given to both the general and goods managers that no rebates were to be offered to any of Wantage's other coal merchants. The directors would be pleased to consider any applications for a rebate for the new financial year which had just begun at the end of the year on 1 October 1883. In the meantime, the managers were instructed to inform the traders that they should write to the company requesting a rebate which may be considered at that time and if granted would be paid retrospectively.

The company received another claim for damages relating to goods in early November. A Mr. Bennett claimed that a rug, which he claimed was worth £10, had been lost in transit by the tramway and he required compensation to the full value of his goods. The tramway, upon investigation, found that the rug had been handed over to them by the GWR at Wantage Road, but could find no proof that the item had been delivered to Bennett by the tramway's representative. The consignment note was found in the office, but was unsigned, although marked as delivered. The company settled the case by offering Bennett the sum of £5, which he accepted, and the matter was closed. Another parcel had gone missing in August although the company had delayed replying to the claim made by a Mr. Houtton for £1 1s 6d until they had ascertained what might have happened to it. No record appeared to exist for the tramway having received the parcel and Houtton had stopped chasing for recompense so the matter was left unresolved.

The Hughes engine was able to assist No. 5 with the goods traffic and No. 6 with passenger traffic, meaning the Grantham tram could be used selectively as the year drew to a close. The Grantham tram was still being used to complement the passenger services despite the Matthews and Hughes tram engines covering most of the traffic. The use of the Grantham tram seemed to remain relatively frequent as the company purchased a waterproof cape on 15 November for its driver to use in the unprotected cab when the weather was bad.

The tramway had just got all of its locomotives back into working order when an accident occurred which damaged both No. 5 and the Grantham Car in early December. The company engineer Wilshire, who also drove the locomotives when needed left the tramway yard at Wantage with the goods engine without the required authority from the general manager or his nominated deputy, and with insufficient time to reach the loop at Grove Canal Bridge, before the tram from Wantage Road which was to leave there at ten o-clock. The two collided between the canal loop and Wantage Town, damaging both the goods engine and the steam tram. Wilshire was reprimanded by the general manager and the directors on 7 December. All of the tramway's drivers, including Wilshire, were reminded that they were not to leave Wantage with engine, car or train without the knowledge of the duty manager, and only then after receiving clear permission to proceed from the same person. Repairs to the locomotive and tram did not take long, both returning to traffic after a few days, although during that time the Matthews and Hughes tram engines had to work all of the tramway's traffic.

The new year started with the company buying a large two-faced clock for the station at Wantage which was to be used both by passengers awaiting tram services but also by the duty manager to note the time of arrivals and departures at Wantage. The use of one clock by everyone despatching and receiving trains had been encouraged by Aldworth to make the entries in the log book consistent.

The tramway was involved in an incident during late January which also resulted in the resignation of Llewellyn Jotcham as a director of the company. The incident resulted in injury reported as having happened to a Mr. Swiss, although it seems that he soon recovered from his experience. The tram service was travelling along the line, when the driver reported that Swiss had run across the line directly in front of the locomotive, which unable to stop in time, had struck him. Fortunately, his injuries were not too serious, although he still approached the company for compensation despite the accident clearly being his own fault. The directors, despite the tramway having no case to answer, offered to pay Swiss 30s towards his medical costs whilst reiterating that the company accepted no liability whatsoever for the accident. Jotcham resigned shortly afterwards as he believed that his role as a director of the company could cause a conflict of interest with regard to his duties as the district coroner. His resignation was accepted with understanding but regret.

The highways board approached the tramway company again in early February. In a letter to Aldworth, the board complained once more about the height of the rails on the tramway's line next to the main road even though the previous complaint had by then been dealt with and the rails lowered by the tramway's engineer. The complaint was passed on to the directors, with Edward Ormond arranging a meeting with the committee to discuss their demands further on 14 March. At the meeting, the highways committee also complained that, in their opinion, the drains still needed to be improved. The tramway representatives offered, as a gesture of goodwill, to undertake further work on the drains using a truckload of broken pipes to be purchased from the GWR.

On the subject of their complaint that the rails were too high above the level of the road, the highways committee informed those in attendance from the tramway that they were not content with the line 'near the common' since one rail was very slightly above the level of the tarmacadam. The tramway, rather than go to the expense of relaying the line to lower it a very small amount, offered as another gesture of goodwill, to provide the highways committee with twenty tons of Beauchamp Stones which could be used to raise the level of the road, then be covered with a new layer of macadam.

The spring saw the company look to improve the yard at Wantage which had often become boggy and partly flooded during heavy rains over the previous winter. It was decided to bring in a large quantity of gravel and used ballast which would be spread over the yard to improve its drainage. The company opted not to use the railway to bring the aggregate to Wantage Road, but rather to place the traffic with the Wilts and Berks canal since its carriage charges would be lower than the GWR's costs. Forty tons of gravel and ballast were duly delivered by the canal on 1 May at a cost of 4s 3d per ton, with work immediately starting on the resurfacing of the tramway's yard.

Despite the continually improving profitability of the Wantage Tramway's services, the company's finances had been affected heavily by the expenses involved with the repairs to all three locomotives during the year. The impact of the cost of repairing collision damage to No. 5 and the Grantham Tram added to the drain on the tramway's bank account. The shareholders complained about the great cost of repairs at the half yearly meeting when the six-month statement was presented to them on 8 May 1884. The shareholders raised a number of matters related to the management of the line and the work of the engines, expressing concern about how the tramway had been run so far that financial year.

The general feeling of dissatisfaction in the meeting subsided once the directors had explained why the expensive repairs to Nos. 4 and 5 had been necessary as well as the repairs to No. 6 which could not have been foreseen on such a new locomotive. Now that repairs had been completed, the directors said, the Matthews and Hughes trams could be expected to run with only regular maintenance being required, whilst the goods engine was effectively brand new after the investment in the work that had been performed by the GWR at Swindon. It could be expected that all three engines would remain operational for a the next few years, other than during periodic light overhauls. The future of the Grantham Tram remained undecided. Its potential use was limited by its lack of capacity, although it was a useful complement to the longer trains that could be hauled by the two tram engines, and if necessary the goods engine, but no decision had yet been made.

The same month, Davey applied to the tramway for a rebate of £5 against the carriage charges he had paid in connection with his coal traffic already carried in the current financial year. In accordance with the instructions which had been given to Aldworth, the general manager referred the request to the directors. The request was acknowledged and in principle accepted. Davey was informed in a letter from Edward Ormond on 22 May, that the payment had been agreed but would not be paid until the company's final accounts for the year had been made up. The company reserved the right not to pay the rebate if the profits for the year would not allow it.

The tramway's line had become due for its annual inspection by the directors by late May. The directors, general manager and engineer walked the line on 29 May inspecting timbers and rails as well as the canal bridge. The worst sections of the line had been previously inspected by the engineer, who pointed out where the rails and sleepers were in need of replacement as the party progressed down the line. At the end of the day, the directors authorised Wilshire to carry out the work on the line where needed. The bridge was found to be in generally good condition but had become a little rusty.

Wilshire, with the permission of the directors, approached the GWR's engineers' office to request advice on how best to complete the repairs to the line. The GWR's Road Superintendent at Swindon, a Mr. Lodge, agreed to come to Wantage to assist the tramway with recommendations for their renewal programme. Lodge came to the tramway on 5 June and, accompanied by a group of the company's directors, walked the length of the line with Wilshire. The GWR engineer recommended that the rails should be replaced at the places where Wilshire had noted a need for remedial work, but that it would not be necessary to replace most of the sleepers as they were still sound. Lodge recommended that instead of being changed, the sleepers only needed repacking with new ballast. He agreed that the canal bridge at Grove was in good condition, but that it should be scraped to remove all the rust before being repainted. If that were done, the GWR's engineer believed that the bridge was good for years of service.

The summer provided the engineer with more work. The Matthews engine failed despite the comments which had been made to the shareholders about its condition. The failure was caused by an unexpected problem with the locomotive's boiler. The boiler and steam condenser needed patching, which Wilshire was unable to undertake himself, although the work could be done at Wantage if a capable boilersmith could be found. The locomotive required additional work to the grate and safety valve bonnet, but Wilshire and his assistant could undertake that work themselves in the shed at Wantage. The company authorised Wilshire to find a boilersmith to undertake the work to the Matthews engine's boiler, and to employ him on a short-term contract. This was done and work, expected to take two weeks and cost £15, began on 12 June.

At the end of the month, the tramway had another repair to pay for. The boiler of the Grantham Steam Tram was becoming worn out and in need of heavy repair. Swindon were asked if they would be prepared to take on the work, which they agreed to do. The Matthews tram engine was returned to traffic in late June with the boiler repairs completed on time, once tested and back in service the locomotive was able to resume hauling the passenger services. The Grantham Tram, now not being relied upon by the tramway to back up the Hughes engine on the passenger trains, was loaded onto a wagon and sent to Swindon on 26 June. The bad news about the tramway's locomotive fleet didn't end with the despatch of the Grantham tram to the GWR's workshops. On 10 July, the goods engine was reported to have broken a driving crank pin, which rendered the locomotive inoperable. The repair was beyond the capabilities of the workshop in the locomotive shed at Wantage, and No. 5 had to be sent to Swindon for the crank pin to be replaced. The locomotive returned to Wantage in the late autumn.

The company looked to divest itself of redundant equipment as the financial year drew to a close. The omnibus which had been purchased from Thomas Nunney to cover passenger services until the second tram car had arrived on the line from the Starbuck Car and Wagon Company had remained little used since then. The bus had remained in the sheds at Wantage even when the tramway was at the height of its locomotive crisis with only the Hughes tram engine and the Grantham Car operational. It was decided to sell the bus as the company had no further use for it at the time, and it was duly sold for £2 on 14 August.

A number of near misses had occurred over the summer where the tramway line crossed over the public road at the bottom end of Grove Street as it left the tramway yard. The manager reminded the company's drivers that they should exercise extreme caution when crossing the road, and added a new requirement that all trams should be carefully brought to a complete halt at the gates of the yard on the edge of the road before proceeding across the road once it was safe to do so.

The company began the new financial year by amending the passenger timetable with the change to take effect from 30 October 1884. It had been found that the tram leaving Wantage Road after the arrival of the GWR's 10:30am train was running almost empty except on Wednesdays. Very few passengers were alighting from the GWR's train intending to travel on the Wantage during the week, the increased volume of passengers on Wednesdays were people going on to Wantage to attend the market for shopping. It was resolved to no longer send a service to meet the 10:30am train apart from on Wednesdays.

The company had received a strong complaint from a passenger earlier in October, which the general manager was asked to investigate. The complainant had alleged that one of the company's servants 'had been insolent'. The servant in question was the conductor of the tram upon which the passenger was intending to travel from Wantage Road. The passenger claimed that when asked to help with their luggage, the conductor had passed comment and refused to help. When interviewed by Aldworth, the conductor admitted that he remembered the incident and that he had indeed been rude to the passenger. The conductor promised not to repeat the offence, and offered to visit the passenger, who lived in the town, to apologise for his rudeness, doing so shortly after his meeting with Aldworth. The company also sent a letter of apology to the passenger.

Chapter 8

CONSOLIDATION, EXPANSION, AND IMPROVEMENT

The new year, 1885, began with the tramway deciding to move the goods manager's office out of the station house and into the goods shed where he would be better placed to oversee the company's freight traffic. A new office was constructed in one end of the shed and fitted out for the manager to move into.

The tramway was busy, with passenger and goods traffic remaining profitable as the spring approached. In March, the company sold its fourth season ticket. A request was received from Edmund Atkins, who asked to purchase an annual season ticket which would allow him to travel on the tramway at any time in second class. The request was granted, with Atkins being informed that the cost would be £7 for the year. Atkins paid the requested season ticket fare, and the pass was duly issued.

The increasing traffic was beginning to worry the engineer by early May, who approached the board to air his concerns. Wilshire believed that the heavy work being undertaken by No. 5 on the goods traffic was causing stress on the locomotive's connecting rods, and he was worried that they might break if strained too much by the weight of the trains and the way in which it might be driven if the train proved difficult to start out of Wantage Road yard. The directors were sympathetic to Wilshire's concerns and authorised him to approach Swindon to have the GWR manufacture a new stronger pair of connecting rods for the locomotive, to replace the existing ones on the engine.

The expansion of the tramway's passenger traffic meant that the company found itself in need of an additional conductor for the trams. Adverts were placed and the company offered William Heedson, aged 17, a one month trial on the tramway beginning on 15 June on a wage of 10s per week. The increased revenue allowed the company to award its most junior staff a pay rise as well as recruiting a new conductor. The 'parcels boy' and 'engineer's boys' were awarded a rise taking their pay from 6s to 7s a week. Meanwhile, the company decided that the facilities at Wantage Town required smartening up. The inside of the station house at Wantage was completely cleaned, and all the offices and the waiting room were then painted

Wantage Tramway Nos. 5 and 7 outside the locomotive shed at Wantage, just before the Second World War. The stronger connecting rods manufactured by the GWR at Swindon can be seen on the engine. *John Scott-Morgan Collection*

with whitewash to make them brighter and more welcoming. As part of the improvements to the station the tramway purchased three gas lamps for the station platform at Wantage for 15s apiece.

1885 had been another good year for the tramway. The company had once again seen increasing levels of goods and passenger traffic. The lucrative goods traffic had added to the revenue for the year, resulting in the company being able to pay a dividend for the year once more. The increased profits allowed the directors of the company to opt to pay £127 6s 1d from the year's profits, with £72 13s 1d added from the reserve fund, to pay off part of the existing mortgages on the company's property. The directors opted to carry over £19 9s 6d, the profit remaining after payment of dividends, to the reserve fund for future repairs to rolling stock and infrastructure. The directors also opted to pay a bonus to Edward Ormond for his work as company secretary, £50 being awarded as a thank you for his efforts. A further gratuity payment was voted for Ormond of an additional £25 for his services over the past year, and £10 each for James Clarke and John Mackay for their support and work during 1885.

Once these payments had been made, the company was left with a profit for the financial year ending 1 October 1885 of £498 10s 0d, all of which was allocated for the payment of dividends to shareholders. Each share was to receive a 5 per cent dividend. The company directors opted to clear the mortgages through the issue of 600 new £5 preference shares at a guaranteed 4½ per cent dividend, effectively turning the loans into share capital with the dividends replacing the interest on the original loans at the same rate. Given that the mortgage providers were all shareholders and directors of the company, the conversion of the loans to share capital was not likely to have been voted against as it was in the best interests of both the company and the loan providers. The preference shares were offered to the existing shareholders in proportion to the number of ordinary shares they held. The capital raised through the purchase of the shares would in turn be used to clear the outstanding loans. All of the preference share options were taken up, the loans cleared, and the company becoming debt free.

At the directors' meeting of 30 November 1885, held in Wantage, Edward Ormond and Thomas Bennett were re-elected as company directors, their term of office having come to an end. The Reverend Alfred Herbert was elected as a new director, on the condition that he increased his shareholding to the twenty shares which were required in the company articles of association in order to be a director of the tramway. The directors of the company were now Colonel Robert Loyd Lindsay who had been awarded a peerage and was now Lord Wantage (Chairman), Philip Gibbons, Benjamin Castle, Colonel C.H. Loyd Lindsay, Joseph Lewis, James Clark, John Mackay, The Reverend Alfred Herbert, William Caudwell, and Edward Ormond (Company Secretary).

The new year saw the company mainly consolidating its operations, only minor changes being made. The comfort of passengers was improved in January when the company ordered cushions to be placed on the seats of one of the tramway's passenger carrying vehicles. Which of the company's vehicles was to receive these cushions is unclear, the only references to it being frequent mentions of 'the tram'. The term had previously been used by the company in its internal correspondence to refer to the Grantham Patent Steam Tram, so it seems likely that the improvements in comfort were destined for the steam tramcar. The other hauled tramcars had been built with more luxury in their interiors which also suggests that the company was trying to improve the customer experience when travelling on the Grantham Tram. The only other major maintenance seen that year was performed on the two passenger cars, both of which were revarnished in June. A bid from Johnsons of Wantage to undertake the work in response to the tramway's tender was accepted for a cost of £3.

Little else happened during the year apart from the annual inspection by the company directors which was completed on Saturday 4 December 1886. Nothing of note seems to have been found as no report was made at the next directors' meeting in January 1887.

After a quiet year in 1886, 1887 proved more eventful. The tramway was involved in an accident on 7 January when a boy, the son of John Saunders was injured when he fell off the Grantham Tram which was being driven quickly from Wantage Road station. Saunders approached the company for recompense for the medical expenses associated with the child's injuries and for compensation for the accident which he believed the tramway had caused by allowing the tram to be driven too quickly. The company directors agreed, on 3 February, that the accident was probably the fault of the tramway and offered to pay Saunders's medical expenses of £1 6s 6d, the manager was also instructed to settle Saunders's claim for damages, paying 12s.

The incident did not end with the payments, the boy's father approaching the tramway again in late February demanding that the company pay the expense of surgery on the child's foot which had become necessary. This time the company disagreed with the claim, replying that the directors believed that the doctor's fee of 1 guinea for a minor surgical intervention was excessive and refused to pay any more damages.

The troubles continued in May, when the tramway was accused of losing baggage belonging to Henry Ellington. After investigation, it transpired that the baggage had arrived safely at Wantage Road. Ellington had been travelling on to London and so the claim was referred to the GWR. After a search, the Great Western found the bags at Paddington where they appeared to have been left behind when the delivery cart left the station. The luggage was returned to its owner, with the GWR offering a small payment as an apology and gesture of goodwill.

The tramway company experienced a problem with the canal crossing during the early part of 1887. The bridge over the Wilts and Berks canal was beginning to show signs of weakness which had been observed by the maintenance workmen working on the line. An inspection by the company engineer revealed that the iron work on the bridge had deteriorated more quickly than would have been expected, possibly due to the use of heavier steam locomotives on the tramway. It was decided at a meeting of the company directors that a new bridge should be constructed to replace the original before it became unsafe.

The company approached Swindon for help, and the company, as willing to help as ever, arranged for the district engineer to meet George Stevenson at Wantage Road on 2 June to examine the bridge and advise on the work that was needed to return it to good order. Whilst at Wantage Road, Stevenson and the GWR's engineer took the opportunity to discuss the layout of the tramway's facilities at the station. Both men examined the rails which led into the passenger terminus platform, which the tramway had wanted to move closer to the platform to allow passengers to board and alight from the tram car more easily and safely. The Great Western's engineer informed Stevenson of the best alignment that would achieve the tramway's needs and promised to cost for the work to be done by the GWR as soon as he returned to Swindon, and to provide a more detailed description of the work that needed to be done to repair the canal bridge.

The GWR engineer wrote, as promised, advising the tramway on the work needed to strengthen the Grove Canal Bridge, and also the suggested new alignment of the line and run round loop in the tramway's passenger terminal at Wantage Road. The proposal was accepted by the board and the GWR undertook the relaying of the line and loop. The Great Western's engineer informed Stevenson that upon reflection he did not recommend that the tramway add a level crossing at the entrance to the yard. It was viewed as both an unnecessary expense for the tramway and could cause confusion to passengers who might attempt to use it to access both the tramway's and the GWR's services instead of accessing the railways through the station building.

The tramway, meanwhile, was becoming increasingly busy, and it was decided that the goods engine driver needed an assistant to undertake some of his work. Adverts were, as usual, placed in local papers. On 2 June, Robert Greenacre was appointed to the role on the wage of eleven shillings per week. The company received sad news when they were informed that their passenger driver, Wiltshire (not to be confused with Wilshire the company's engineer) had died. Alfred Futcher was employed to take over the duties of passenger driver at a wage of 16s per week.

The requirement for the replacement bridge structure was put out to tender, with Alfred Anderton and Sons of Accrington, Lancashire submitting the winning bid. The company was a long-established firm of engineers and boiler makers, and the structure was soon delivered to Wantage Road station yard by rail. Anderton provided an engineer and workforce to install the bridge. The works were undertaken during August and the 44½ft long skew bridge was soon in place. Whilst the bridge was being replaced the tramway company took the opportunity to relay the curves on the approaches to the bridge to make them less sharp. The new layout of the line removed a problem with the original design of the route, the more open curves allowing the trams to continue at line speed through the approaches rather than slowing to negotiate the sharp bends. The line was also levelled and the ballast replaced during the works, improving the ride for the tramway's passengers greatly.

In accordance with the requirements of the BoT, Edward Ormond invited the board to send an inspector to survey the new bridge to certify its fitness for use. The BoT once again sent Colonel Hutchinson to Wantage to perform the inspection. Hutchinson arrived at Wantage Road to perform the inspection on 15 October, but all was not good. Hutchinson found that the realignment of the lines approaching the bridge had been accomplished to a good standard but there was a structural concern involving the bridge itself.

Colonel Hutchinson's report stated that the girders used in the construction of the bridge were not all acceptable. He had found that one cross girder and one angle iron strut had needlessly been pierced with rivet holes which remained empty. The holes had weakened the structure and Colonel Hutchinson was thus unwilling to pass the bridge as safe until remedial works had been completed.

Despite being referred to as 'The Goods Engine' by the company, the tramway often used No. 7 on passenger services. Here the locomotive is working the service with just car 4. The tight curvature of the tramway can clearly be seen behind the tram as the line follows the road. The close proximity of tramway to roadway was partly the cause of the damage done by the United States Army Air Force in 1944. The rod pointing out from under the rear coupling is the tramcar coupling pole. *John Scott-Morgan Collection*

Edward Ormond informed Alfred Anderton of the issue with the bridge and he devised a solution to the problem. Ormond wrote to the BoT on 9 November to inform them that the problem had been remedied and that the bridge was now ready for a second inspection. In the letter, Ormond outlined the work that had been undertaken to improve the structure of the canal bridge. A set of additional flange plates had been rivetted onto the pierced girder and angle iron strut. The flanges provided additional strength whilst the rivets had filled the drilled holes. Colonel Hutchinson once again inspected the bridge, asking for the tram locomotive to be run back and forth across the bridge and then for it to be stopped on the centre of the span. The bridge performed admirably and Hutchinson provided a report allowing the tramway to restart services across the canal. The Colonel suggested one optional improvement to the bridge, advising that it might be wise to cover the wooden floor between the rails with a thin layer of ballast to prevent sparks from the engines setting the timbers alight. It seems that Colonel Hutchinson was willing to be quite helpful to the tramway, this not being the first time he had used his experience from his time as an officer in the Royal Engineers to help the company.

1887 continued to be eventful right until the end of the year. The tramway was involved in another accident in October. The Grantham Car, travelling along the line at speed, suddenly braked heavily throwing some of its passengers to the floor, whilst others were thrown off the tram completely. At the 3 November directors' meeting, the board discussed the incident with both the general manager and driver Futcher, who had been driving the Grantham Car at the time. It was agreed that the driver had behaved correctly in braking to avoid a collision, and that no liability could be accepted on behalf of the tramway by either the manager or the board.

The new bridge over the canal was performing well after the rework had been completed by Andertons. The old bridge, stored at Wantage, was sold to the Wilts and Berks Canal Company for £30 on 5 January. The canal company had only needed the tie rods from the bridge and once they had been removed, the remains of the bridge were sold to the Cardiff Iron Company for £28 as scrap.

The company sought to improve the pay it was offering its staff and contractors during the year. In May, Ryman, the goods agent, was offered commission on the monies he was collecting for delivery charges when dropping off people's goods. The commission was agreed at 3d per £1 collected on the tramway's behalf on 3 May. At the same time, the directors instructed Ryman that no deliveries were to be made to any consignees without payment of the tramway's charges unless they already had an account with the company, as a minority of individuals had refused to pay the tramway once delivery had been made. More pay rises followed in July with Harry Hudson, the passenger tram conductor receiving a rise from 12 to 13s a week. At the same time, the goods driver's assistant, Robert Greenacre, and Hedley, the 'engineer's boy' (junior assistant) were both given a raise from 6 to 7s a week.

The goods engine was sent to Swindon for new tyres to be fitted to its wheels in August, and whilst there the locomotive underwent a light overhaul which included a repaint into GWR lined green, although carrying W.T.C. on the cab side instead of the GWR's preferred statement of ownership at the time. The tramway's motive power received a permanent boost just after No. 5 returned to Wantage. On 4 October, the tramway took up the option offered to them by the owner and purchased the Matthews tram from them for just £60.

As the calendar year drew to a close the tramway was approached with an offer of new traffic for the line. The company was asked if they would be willing to carry the evening papers off the GWR's early evening services from London on their passenger cars from Wantage Road station to the town. The tramway offered to carry the newspapers for a charge of £5 a year, which was agreed on 20 December.

Wantage's evening newspapers continued to be delivered to the town by the tramway until the end, the last goods train bringing the papers in one of the vans. At the same directors' meeting, it was agreed to do away with having first and second class in the passenger cars, instead one simple fare would be charged for all passenger journeys going forward.

The company continued to be successful in 1888. The accounts for the first half year between 1 October 1887 and 31 March 1888 showed a good profit. The company's takings for passenger traffic had risen to £297 9s 3d, with goods and parcels traffic receipts increasing to £1,271 8s 8d and £288 11s 3d respectively. The company had paid the GWR £684 2s out of the goods revenue for charges it had collected on the railway's behalf for onward carriage of goods from Wantage Road, with £197 14s 4d of the parcels revenue likewise passed on to the Great Western. Wages had increased slightly to £288 15s 0d for the tramway company's staff. The company was left with a good profit and a dividend was declared by the directors for the full amount due on the preference shares for the previous financial year ending 1 October 1887 amounting to £65 10s 8d. The ordinary shareholders had been granted a dividend at the usual rate of 4 per cent for the 1887 financial year, amounting to a total payment of £326 16s. The tramway had once again proved to be a safe investment during 1887.

The second half of the year saw good returns on passenger, goods and parcels traffic, with a bonus to the company's finances from the sale of the old canal bridge which had been sold for scrap after its replacement with a new crossing. The company finished the year with a profit of £445 5s 9d after all expenses, including the payment of preference share dividends and taxes. Once again, the directors declared a dividend of 4 per cent on ordinary shares, the shareholders receiving another good return on their investment.

The company had been paying a 'quit rent', a land tax imposed of owners of freehold property in lieu of services to local landowners. The system was one of the last vestiges of the old feudal system which imposed a fee of 5d payable to the Lords of the Manor of Wantage related to the tramway's property at Wantage. The tramway directors resolved to establish whether the Lords of the Manor were willing to sell the rights to the quit rent to the company, with Edward Ormond being instructed to write to the 'Steward of the Manor' who represented the Lords to establish if they would sell the rights. The Steward replied that they would be prepared to sell the rights for the sum of 10s 5d. The directors duly agreed to do so and the company secretary, Edward Ormond, wrote a confirmation letter accompanied by a cheque for the agreed amount which was sent to the Steward of the Manor, Edward Ormond! The correspondence where Edward Ormond effectively wrote to himself still survives in the Wantage Tramway collection in the Royal Berkshire Archives in Reading.

The first thing that the tramway did in January 1889 was to replace the doors in the dividing wall between first and second class on their passenger cars which had been removed to make access easier for the conductor after the abandonment of separate classes on the tramway. Some passengers had complained about the draught through the centre of the cars, so the company decided to replace the doors.

The company had nearly been involved in a potentially serious accident in the last few days of 1888, which was still to be investigated by the company and the local police and was discussed at the board meeting with the general and goods managers also in attendance. The incident had occurred on Friday 28 December, and was referred by the police as a 'malicious act'. An unknown person had thrown a large iron bar across the tramway, just before the passenger engine left Wantage Town to meet the GWR's 5.55pm train. It had clearly been an attempt to derail and damage the tram and may well have resulted in serious injuries to the driver and the passengers on board the service. Fortunately, the driver was keeping a good lookout and spotted the obstruction in plenty of time to bring the tram to a standstill before leaving his engine and removing the obstacle. It was reported when the tram reached Wantage Road.

The tramway had been carrying an increasing amount of traffic for the Wantage Gas Works, both coal in to the works, and coke, as well as the two byproducts of the process of producing town gas, out of the town. Large quantities of coke were transported over the tramway alongside a thick tar which was used for road surfaces, the tar in tarmacadam, and a pungent liquid. The liquid, referred to as 'ammoniacal liquor' was carbolic acid, which was used to make disinfectants and antiseptic soap. So much of the latter was being carried by the tramway, that the company informed the Wantage Gas Works Company that they were reducing their carriage charge by half from 2s to 1s per ton of the liquid.

The goods driver's assistant, Robert Greenacre was promoted to full driver on 6 June 1889, and his wages were increased to 13s per week. The same month saw the tramway begin to consider the future of the Grantham Steam Tram. The directors decided that in the long term, the steam machinery was to be removed from the steam tram, and after that had been done, the car should be returned to service as an ordinary car to be hauled by the passenger engine. In the short term, the Grantham tram remained in service until the time came to remove the machinery.

The tramway line was becoming in need of renewal towards the end of 1889, and in early 1890 it was decided to replace the permanent way with new rails and whilst doing so to dispense with the longitudinal baulks supporting the rails and to replace them with the more usual transverse sleepers. The tramway wrote to the GWR Chief Storekeeper's Office at Swindon on 9 January to enquire if they would be able to supply the required number of sleepers from their redundant used stock. The GWR Storekeeper was willing to sell the tramway what they needed, and an order was placed with the Great Western for five tons of used timber sleepers. The sleepers were no longer suitable for the heavy weights and high speeds of trains on the GWR but were more than sufficient for the lighter and slower traffic on the

Wantage Tramway. The tramway company directors once again used their close relationship with the GWR to recycle materials, saving the company money.

The passenger service had become increasingly busy in the latter part of 1889, and with passenger traffic continuing to increase, the tramway company made the decision to employ another passenger tram conductor. Following the publication of the vacancy in the local newspapers, and a series of short interviews, John Harris was appointed as the company's new conductor on a wage of twelve shillings per week.

The Grantham Steam Tram, already the subject of debate among the tramway's management, broke down in March. The vehicle had developed numerous steam leaks, and had broken one of its four wheels. The decision was made on 6 March to repair the steam leaks in the yard at Wantage, the company's engineer was instructed to carry out the repairs. The broken wheel was to be replaced with a cast iron wheel made at the Vale of the White Horse Foundry as a temporary fix. The Grantham Tram was then to be taken out of traffic but stored where it would not deteriorate in case it was needed. The steam tram did not run on the tramway again and eventually the machinery was removed and scrapped.

With the failure of the Grantham tram and its removal from regular service, the tramway needed a new passenger car to lengthen its passenger trains. Discussions were already underway with a selection of tram car builders prior to the failure of the steam tram. It was decided, at the same time that the Grantham tram was withdrawn, that the company would purchase a new car from Traceway Carriage and Wagon Builders in Birkenhead since their price was acceptable. One modification was requested on the new passenger car, the seat backs were to be fitted with cushions. The builders agreed to do so, and the car was ordered at a cost of £175 on 3 April.

The tram station at Wantage was in need of rebuilding by the beginning of 1890 and became progressively worse during the first half of the year, with the platform roof especially requiring heavy maintenance. The roof over the platform had begun to leak through broken or missing slates. The company approached builders in Wantage for an estimate for the work required to make good the tram station, with Joseph Aldworth being awarded the work in response to his 24 June estimate of £8 15s to strip the roof, replace the roof board and recover it with new slates.

Goods traffic was increasingly busy over the course of the year, and the goods driver, Robert Greenacre, was working longer hours. The directors recognised the extra effort that their driver was devoting to the company and awarded him another pay rise taking his wages to 14s per week. The company received some disappointing employee news in late September, when the company's general manager, Aldworth, gave notice of his intention to leave the company. Announcements were made in the local press, and national engineering journals, that the tramway was seeking a new manager on 2 October. The company's search for a replacement manager remained unresolved until the very end of the year, the directors reporting in their

6 November meeting that, whilst many candidates had applied for the position, with some being interviewed for the post, none of them were found to be suitable. The company's search for a new manager finally ended when they appointed a new man to the role, with William Watson taking up the position on 1 January 1891.

With the new year came a need to invest once more in renewals of the tramway line. Parts of the line were in need of repacking and where water had been collecting during the early part of the winter, the rails needed to be raised to prevent the large puddles from forming. The work would be expensive and was the reason that when Hewitt, the passenger driver, and Spicer, one of the two goods drivers, approached the directors with a request for their wages to be raised from 22 to 24s a week, their request was denied. The company directors, despite declining applications for pay increases, offered their employees a new benefit. From 2 July 1891, all of the tramway staff were entitled to free travel on the line at any time on production of their employee paycheque tokens. Another benefit followed shortly afterwards, when the tramway's employees were informed that, as of 9 July, the company directors had taken out an insurance policy, at no cost to the staff, which would cover them in case of accident whilst working on or about the line. The policy would cover the pay of the employee whilst unable to work and, should the worst happen, would pay a pension to their family. The insurance did not cover the tramway's employees for loss of earnings whilst off work with an illness. Statutory sick pay was not yet available, the necessary legislation still being a long time in the future.

In November, James Clark, the owner of the flour mill in Mill Street, Wantage, approached the tramway company requesting reduced rates for conveyance of corn and flour over the tramway. Since the mill was sending a very large amount of both commodities over the line, the directors authorised the manager to grant a reduction in charges, on the condition that the mill placed at least a combined 2,000 tons of corn and flour per year with the tramway for conveyance to Wantage Road. The normal carriage rate for the type of freight that the tramway was carrying on behalf of the mill was 1s 3d a ton. If the mill transported 2,000 tons during the year, the rates would be reduced to 1s 1d per ton, if 2,500 tons or more were carried, the rate to Clarks Mill would be 1s. If the conditions were met, the tramway would pay the difference as a rebate to Clark at the end of the financial year.

At the end of the year, one of the company's conductors, Harris, informed the company that he was leaving the tramway to take up better paid employment elsewhere. The search for a new conductor began with the company placing adverts for the vacancy with local newspapers.

The new year began with the vacancy for the replacement conductor being filled. The company appointed William Savory, a 31-year-old local man, to the role on a wage of 14s a week. Savory was to take up his new post from 7 January. On the same day, passenger car No. 3 had to be withdrawn from service for repairs. The car had been damaged whilst in transit on the line by one of its passengers, making the car unsuitable for continued use. The passenger accepted liability for the damage that he had caused and offered to pay for the repairs.

At the end of 1891, Wilshire, the company's engineer, had been taken ill and was off work for some time suffering with influenza. The company, had as was normal practice at the time, stopped his pay whilst he was unable to attend work. Upon his return, Wilshire wrote a letter to the directors, politely requesting whether the company might see fit to pay his weekly wages for the month that he had been absent through his illness. Since he often spent a large amount of his own spare time in the locomotive sheds or out on the line for the benefit of the tram company, the directors agreed unanimously that their engineer more than deserved to be cared for by the company, and as such the wages which had been stopped during his absence would be paid as a bonus alongside his next weekly wage.

The repairs to the tramway had been completed during the spring, the last major piece of work was finished, which resolved a long-standing flood risk on the curve at the approach to the southside of the canal bridge where water accummulated in bad weather. The drainage ditch next to the tramway by the curve was often blocked by a build up of fallen leaves and mud, which caused water to overflow onto the line. The tramway filled the ditch with old timbers purchased cheaply from the GWR, with a drain made from old pipes, also purchased from the railway company, used to form a culvert at the bottom of the ditch. The work prevented the ditch filling up and let water flow away from the site.

The completion of the works on the line allowed the directors to revisit the wages and salaries that were being paid to the tramway's staff. The conductor's wages were all increased to 15s per week (£39 per year), and the manager's salary was raised by £10 per year to £110 on 1 July. Remembering John Hewitt, the passenger driver, had requested a pay increase at a previous meeting, the directors awarded him the rise that he had requested. Hewitt's wages were increased by 2s a week to 24s (£62.8s per year) from 4 August.

The delivery of goods seems to have still been causing the company a problem. Some consignees were claiming that their goods had not been delivered or had arrived damaged. In order to prevent future spurious claims, James Sims informed the goods delivery agent that from 1 September, he was no longer to deliver any goods unless the consignee or a representative had signed the delivery sheet acknowledging receipt of their goods. The regulation appeared to resolve the problem, the company no longer receiving fraudulent claims related to deliveries.

October saw the company presented with the need to undertake more repairs on the line, as the last of the rails which had originally been used in construction of the line continued to deteriorate. The company took the decision to replace the lighter original rails with more robust heavier steel rails. Fifteen tons of flange rails were purchased from a foundry in London for the sum of £4 5s per ton, with fish plates at a further £8 10s per ton. The rails were due to be delivered to Wantage Road at the end of the year.

Repairs to the line were not the only maintenance requirement which was occupying the thoughts of engineer Wilshire. The company had three locomotives in traffic, the Matthews tram engine, No. 6, hauling the line's passenger traffic, whilst the goods engine 'Jane', No. 5, was busily working away on the line's freight services. The Hughes tram engine, No. 4, was sharing both the passenger and goods traffic with the other two engines. The Grantham steam tram was still laid up out of use at the back of the yard in Wantage. The two tram engines were beginning to show signs of wear and were becoming in need of an overhaul, but both could not be away from the line at the same time. It was decided on 6 October that the Matthews tram engine, which was less worn out, should be temporarily repaired by the company engineer in the tramway's workshop at Wantage so that the more mechanically tired Hughes tram engine could be sent to Swindon for a major overhaul.

The rails which had been ordered by the company the previous October had been delivered to the GWR's yard at Wantage Road very early in the new year, but there they remained. The rails had been made to the wrong section and were unusable. Edward Ormond wrote to the manufacturer on 5 January informing them that the rails were unsuitable and not as ordered. The company could therefore not accept them, and they remained in the care of the GWR, from where they would deliver them, at the manufacturer's cost, to anywhere he chose to direct them. The tramway company turned once more to their allies at Swindon and purchased used rails from the GWR which were used to improve the stretches of their line where it was needed.

January 1893 saw the tramway involved in an accident. On 18 January, R.P. Langford was driving his wagonette along the road alongside the tramway when one of the line's engines approached. The presence of the locomotive caused the skittish horse to bolt, Langford losing control of both horse and wagon. The driver of the cart leapt from his out of control vehicle, fell and broke his leg. Langford wrote to the company's general manager blaming the tramway for his misfortune and demanding compensation. The manager responded, quite rightly denying any liability for the accident on the behalf of both the company and all of its locomotive drivers.

The spring arrived and with it increasing goods traffic on the tramway. The Hughes engine, now back from Swindon and in traffic was proving unsuitable for the increased length of the goods trains now required. The need for the Hughes locomotive to complete two journeys for the same loads that No. 5 could handle in just a single trip led the company to revisit its motive power fleet. On 6 April the line's manager was instructed to locate a suitable tank engine for the goods work, and to begin by approaching both the GWR at Swindon, and the London and North Western Railway at Crewe to ask if either company had a modern second hand locomotive which they would be prepared to sell to the Wantage Tramway. The manager was also instructed to place an advertisement in 'the engineering newspaper' stating the tramway's need and the specification of the locomotive required.

Later in 1893 the company was involved in another serious accident. The tram, noted in reports as being a steam locomotive and tramcar, collided with a horse and cart on 15 April. At that date the steam engine would most likely have been the Matthews Tram locomotive No. 6. The accident happened when the horse and cart, being driven by Edmund Rivers, emerged from the lodge entrance to Lord Wantage's estate directly in front of the tram. Rivers was travelling along the road in Lord Wantage's estate against Wantage's request, having been asked to use a different exit from the estate, and was warned about being careful when crossing the tramway. The cart drove straight across the rails, despite the tram sounding its whistle as it approached the road crossing, giving the driver of the tram locomotive absolutely no opportunity to stop the train before it hit the cart. Edmund Rivers was thrown from the cart and killed.

The inquest, held on 4 May, exonerated the driver of the locomotive. The coroner gave a ruling of accidental death, with no blame being attributed to either the driver of the tram or the tramway. The manager had, some considerable time before the accident, instructed the tramway's drivers to sound their engine's whistles at the gas works in the yard at Wantage and as they approached the crossing over the public road at Wantage, the entrance to Greyfriars Farm, either end of the canal bridge, the entrance to the Saw Mills, Windsor Lane, Lord Wantage's Lodge entrance, and at the Volunteer's Inn where the line entered the GWR's yard at Wantage Road. The driver had been following the instruction to the letter on the day of the tragic accident, but the requirement was reiterated to all of the tramway's drivers following the inquest.

The company had received responses from its enquiries about the purchase of a goods engine by May. The GWR had responded that they regrettably did not have a suitable locomotive available at the time, the London and North Western responded in a similar manner. One company had contacted the tramway with an offer. The contractors who had built the Manchester Ship Canal was disposing of the equipment used on the canal now that the project had been completed. Among the plant that was being sold were a few tank locomotives, which the company was invited to send representatives to inspect. The meeting was scheduled for 1 June 1893.

Clark, representing the board of directors and Wilshire, as company engineer, travelled to Manchester to inspect the locomotives that the contractor had in store pending sale. The two men were informed that two of the engines were not currently available as they were worn out and required overhaul. Once the overhaul had been completed, the two engines were to be sold by auction, although the other locomotives were all available immediately if the company was interested in making an offer for any of them. The tramway's representatives inspected all of the locomotives, and found that two, named *Brassey* and *Seedley*, would be ideal for their needs. A price was requested from the contractor before the two men set off on their train journey back to Wantage.

The tramway received a letter from Thomas Walker and Company, the owners of the pair of locomotives in which the prices for each of the two locomotives and the conditions of sale were provided. *Brassey* was offered to the tramway company for the sum of £330, whilst *Seedley* would cost the Wantage Tramway Company £320. Both locomotives were sold as they stood in the sheds at Salford Docks. The tramway opted to purchase *Brassey*, but offered £300 for the locomotive, provided that Walker delivered it to the GWR at Manchester. The offer was accepted and the tramway had its new engine.

A letter was unexpectedly received by the tramway's manager in early August. The solicitors, Messrs. Sedgefield, representing Langford who had fallen from his runaway carriage in January, wrote enquiring what compensation the company intended paying their client. Langford had engaged the solicitors despite the tramway directors having made it clear to him that the company accepted no liability and was in no way at fault for the accident. A reply was sent to the solicitors,

The Matthews engine at Wantage with passenger cars 3 and 5. The passenger cars are notable by being in the later livery with the car numbers removed and the company name brought up to the panels directly below the windows. *John Scott-Morgan Collection*

informing them that in no way did the directors accept any liability nor was the company guilty of any negligence.

Whilst dealing with Langford's claim, the company was confronted with two more incidents to manage. On 3 August, the tramway's general manager reported that the driver of 'Mr. Silver's cart' had pulled it into the side of passenger car No. 3 and had damaged one panel which would need to be repaired. On 2 September another accident occurred, one of the passenger cars derailing with one passenger, Mrs. Pottinger, being lightly injured. The company offered to pay £2 compensation to Pottinger in view of her injuries.

The disagreement with Langford continued. Messrs Sedgefield wrote back to the tramway on 30 November asking the directors to reconsider their position and make an offer to their client. The tramway responded stating that the directors had made a full investigation of the incident at the time and they found no evidence of negligence, and so would not be making any offer to Langford. Around the same time, the company received another claim for compensation, this time relating to the derailment of the tramcar in September. E.J. Wallis claimed that his wife had been travelling on the service in the car which derailed. She had been injured with the expenses for her medical care that followed amounting to £2 5s 6d which he wanted the company to pay, in addition to damages of £15, a total of £17 5s 6d. Since the company had settled the claim from Pottinger for £2, they believed that they ought to settle Wallis's claim, but for no more than £5, although they were prepared to pay the medical bills. The company wrote to Wallis accordingly.

The year ended with the company putting out tenders for the construction of an extension to the locomotive shed at Wantage. A bid from Greens, a builder, for £50 was accepted on 14 December, with the condition that the timbers used in the construction of the building were 'specified strongly'. At the same time, the general and goods managers were instructed not to let any of the company's customers be indebted to the company for more than one month, which seems to suggest that the tramway's customers were once again a little reticent to pay their bills on time.

As 1893 turned into 1894, the company had been discussing the erection of the workshop extension with the contractor. Greens had submitted a revised estimate for the stronger building specified by the company, at a raised cost of £64. The estimate was accepted and the construction of the new home for the tramway's engineers started.

The claims for damages from incidents in the previous year had still not been resolved by January 1896. Wallis had engaged the services of a solicitor, Jewell and Sons, who wrote to the tramway demanding that the company pay the full £17 5s 6d damages that their client had requested in November. The response from the company stated that they were not prepared to pay the full amount claimed as they

No. 7 shunting coal wagons in the upper yard at Wantage. The building is the extended running shed. The new workshop, seen to the left with the small window behind the coaling stage, allowed the company's engineer to perform minor repairs on site. *John Scott-Morgan Collection*

thought it was unreasonable. The company however, wished to act in fairness, the directors said, and offered the sum of £12 5s 6d in final settlement of Wallis's claim.

The new year also saw the tramway adding to its infrastructure and services. The increasing amount of trade goods and parcels being carried by the tramway meant that it was often necessary for the goods manager and his assistant to work later, sorting the goods for delivery. The company decided that in order to improve working conditions in the goods shed, gas lighting would be installed.

A new service was added to the tramway's timetable to complement a change that the Great Western had made to its Wantage Road passenger service. The GWR had introduced a slip coach service on its express which passed Wantage Road at 10.30 each day. The slip coach was a carriage at the end of the train away from the locomotive which had a special coupling which could be operated by a second guard stationed in the slip coach. The guard, by operating the coupling mechanism, could uncouple the vehicle from the rest of the train whilst running at speed. The train would then pull away from the loose coach which would then coast, and by skilled use of the brakes the guard would bring the vehicle to a halt in the required station.

The use of slip coaches meant that additional stops could be added to express trains, without the need for the train to be halted in the station and such a stop had been added for Wantage Road. The addition of the regular slip coach on the 10:30 express suggests that there was a considerable number of passengers who wished to make the journey, to justify the expense of operating the coach. In order to provide an onward service for passengers, the tramway arranged for an extra return trip each day from Wantage, scheduled to meet the coach. It might also have been useful to the GWR for the tramway to have a locomotive waiting at Wantage Road.

If the guard misjudged the distance, the coach could stop short of, or overshoot the station it was supposed to stop at and with no power of its own a locomotive would have to be sent for to put the coach where it was supposed to be. In the case of Wantage Road, no doubt there were occasions when the tram locomotive sent with its train of passenger cars to meet the GWR service, was let out a short way onto the GWR's main line to collect the coach and tow it to the platform. The only other option, in the event of the slip coach missing the platform, would have been to ask Didcot to send an engine from the locomotive depot there just to do the same operation.

The Hughes engine needed a thorough overhaul and in January it started to develop leaks in the firebox, which could temporarily be patched, but would need major work in the near future. It was decided that No. 4 was to be sent away to Swindon, where arrangements had been made for the GWR's workshops to manufacture a new copper firebox and complete the overhaul of the engine. Before the locomotive could be sent away, the tramway needed to make some temporary repairs to No. 5, so that the goods engine could continue running. Whilst the repairs to No. 5 were being completed, the Hughes tram remained at Wantage, being patched up to keep it in service.

Whilst No. 5 was being repaired and No. 4 was being constantly patched up, the 'new' locomotive which the tramway had purchased arrived from Manchester, was given the No. 7 and an unofficial name of 'Mary', its previous 'Brassey' nameplates being removed when it arrived. The locomotive needed some minor repair work before it could be put into service, all of which could be done by Wilshire. The tramway company enquired of the locomotive's original manufacturer, Manning Wardle in Leeds, as to whether it might be possible to purchase the required spares from them, if they were willing to make them. Manning Wardle responded that not only would this be possible, but the parts in question were common components and they had a stock of such which they would be able to send immediately upon the tramway placing an order. This was done, and the parts duly arrived at Wantage Road. 'Mary' was soon in service on the tramway, and the company would now

No. 7 outside the running shed at Wantage in April 1923. The Matthews tram engine can just be seen inside the shed. *John Scott Morgan Collection*

have four operational locomotives once the Hughes tram engine returned from Swindon and repairs to No. 5 had been finished.

Traffic on the tramway had continued to grow and with the need to keep all the locomotives in traffic as much as possible, the staff had been extremely busy, often working extra hours voluntarily to keep on top of the workload. The board had recognised that their employees had been working very hard and were discussing what they could do to reward their staff for their efforts. During the summer, the company received a request from the general manager for an increase in pay, who approached the board to ask for a salary increase from £110 to £118 10s 0d per year. When debating the manager's salary, the directors considered the effort given to the company, and the changes which the manager had made to make the tramway both more efficient and safer for everyone who worked and travelled on the line. The manager's request was however rejected as being unsuitable. The directors awarded the manager a better pay rise on 5 July 1894, increasing his salary to £120 per year, with the increase being backdated to June.

With the tramway exceeding all expectations for the year, by early August it was clear that there were funds available to make improvements to the company's rolling stock. The opportunity was taken to repaint three of the company's passenger cars, and one of its engines, all of which had been completed by early September. The engine was not named in the company's communications, but it seems most likely that the Matthews Tram, No. 6, was the locomotive which received a refresh of its paint. It had been found that both No. 5 and No. 7 were as capable of hauling the passenger cars, although they had needed to be fitted with a special coupling in order to do so. All of the company's locomotives could be used on the passenger trams, but the use of the tram engines was preferred.

An unexpected cost confronted the company during late August when locomotive No. 7, the Manning Wardle engine, ran through a set of points in the yard at Wantage Town stations which were set against it, and was derailed. During the accident, one of the locomotive's axles was bent, and needed to be repaired. The engine was jacked up in the yard at Wantage and the axle dropped out of the frames to be sent to Swindon. The GWR, upon inspection of the damaged axle, advised that it could

Wantage Tramway No. 5 at rest in the yard at Wantage in 1924. The curtain which could be used to close off the side of the cab had been drawn forward suggesting it is a warm day. The bar hanging at an angle from the rear buffer beam is the tram car coupling that the locomotive was fitted with to enable it to haul passenger services when needed. *John Scott-Morgan Collection*

not be repaired, and the tramway authorised Swindon to produce a new axle and press the locomotive's wheels onto it. The work did not take long, and the wheelset was soon back on the locomotive.

The profits for the 1893-94 financial year had produced a sufficient surplus to allow the company to decide to send the Hughes tram engine to Swindon for an overhaul, as it was becoming extremely tired mechanically, and often developed minor faults which needed patching to keep it in service. It was agreed with the workshops at Swindon that the engine would be repaired over the winter and since the tramway had another three locomotives in traffic, a full major overhaul would be undertaken.

The new year began with a communication from Swindon which contained bad news. The workshops had sent an engineer to assess the Hughes tram to ascertain what work was needed. On 31 January, the company directors discussed the engineer's report with regard to the state of the Hughes locomotive that the Great Western had sent to the tramway. No. 4 had been found to be in extremely poor condition when examined by the Great Western's engineers. All of the working parts were found to be very heavily worn, with many of them being well beyond repair.

The GWR had recommended that a programme of complete renewal would be needed, with the locomotive being almost completely rebuilt. Whilst the work would be more expensive, it would result in the tramway effectively receiving back a brand-new locomotive. The tramway company had funds available to pay for the work and authorised the workshops at Swindon to proceed with the renewal of the Hughes engine. The tram was sent to Swindon on 7 February and was disassembled for work to start. While the locomotive was being taken apart by the GWR, the boiler was inspected and more bad news was received by the Wantage Tramway Company. The engine's boiler was found to be worn out and unsafe. As it was beyond repair, the GWR's boiler inspector who had examined No. 4's boiler advised that the only viable option was for the existing unit to be scrapped, with the Great Western manufacturing a replacement. The work was agreed on 7 March, after which point the locomotive which would return to Wantage would be a totally new engine, only the frames and a handful of parts remaining from the machine which had arrived at Swindon.

Whilst the renewal of the Hughes engine was ongoing, the tramway looked to improve other aspects of the line, and it was decided that the Grove Canal Bridge ought to be smartened up with new paint. George Richings of Wantage was asked to undertake the work, using 'good oil colours' (enamel paint) for the sum of £4 5s. The wheels of tram car No. 3 were sent away to Swindon for its tyres to be turned, leaving it fit to run for the foreseeable future once the wheels had been refitted.

The tramway directors and all of its employees were saddened to be informed on 4 April, that their goods agent, Ryman had passed away suddenly. The company immediately engaged his son, Fred Ryman, to take on the role of goods agent on a temporary basis, although the arrangement was made permanent in early May. Fred Ryman was employed on the same terms as his father had been, except that he was not required to collect monies for carriage fees. The relationship between the company and their new goods manager did not start well. Fred Ryman's work was not to the satisfaction of the company's directors, or the goods and general managers, and by September he was told to 'square up his accounts', and was informed in writing that his work was not to the standard required, with collections and deliveries not undertaken either regularly or quickly enough.

During August and September, the tramway company and the post office were in discussion regarding the contract under which the mails were being carried from Wantage Road station. The company wished to increase the charge by a small amount to cover their increased costs, but the General Post Office was unwilling to pay more. By 3 October, the negotiations had stalled, but the company, attempting to get a final agreement from the post office, wrote to the postmaster general stating that they were unwilling to carry the mails for less than £1 per week.

October saw the passenger services becoming busier than they ever had been, although that was beginning to cause an issue. The number of passengers in each of the cars was leaving little room for luggage onboard each service, and something needed to be done to provide more capacity. In order to meet the needs of the passenger service, the general manager was instructed to seek estimates from carriage builders for a luggage van which could be attached to the passenger trains. A tram wagon catalogue was received from Kerr Stuart and Company shortly after the manager had written to them. A luggage truck illustrated in the catalogue was of particular interest and the manager was authorised to negotiate with the manufacturer with a view to purchasing one.

The bills for repairs to the Hughes engine were received in early October, the Great Western's account for work to that point amounting to £135 14s 11d, with a further £23 15s 4d due to Robinson and Auden, the new owners of the Vale of the White Horse Ironworks for additional parts which Swindon were not providing. The engine was back on the tramway late in 1896.

Despite the high level of expenditure on repairs to the locomotives and infrastructure during the 1894-95 financial year, the company made a good profit. After part of the profit was transferred to the company's reserve account, a total of £353 6s 4d was available for payment of dividends. At the annual general meeting on 5 December, the directors were, as a result, able to announce a 3 per cent dividend to the shareholders.

1896 started badly for the company, with an accident occurring on the line, and the goods engine becoming unreliable. In early January, the tram service hit a wagon owned by James Clark, the flour mill owner. Clark claimed compensation of

£1 15s 6d for the damage to the wagon, to which the tramway responded offering to share the damage mutually since Clark's men appeared to have contributed to the accident.

The goods engine which had become in need of an urgent overhaul was failing on a regular basis, which led the directors to debate whether the company would be best served by searching for a new goods engine, with No. 5 being sold for scrap, or whether the locomotive should be repaired. It was decided to send No. 5 to Swindon for repair, although the company also contacted Thomas Walker in Manchester, from whom they had purchased No. 7 to enquire if they had any engines left in stock and if so the prices of each. An estimate of £262 was received from the GWR for the repairs to the goods engine on 5 May, shortly after it had arrived. The tramway authorised the work and seem to have ended their interest in purchasing an additional engine from Walker's.

The claim for damage to Clark's wagon in January continued to trouble the tramway in May. Clark was still asking the company to pay the full amount he had claimed. The company responded, asking Clark to settle his claim for £1, but

A study of No. 7 by the wall of the running shed in Wantage upper yard showing the complicated coupling hook arrangement fitted to the engine. The smaller hook was used for fly shunting in the yards. A chain was linked from there to the coupling hook on a wagon and used to move the wagon. The constrained nature of both yards at Wantage made it necessary to work in that manner. *John Scott-Morgan Collection*

also reminded him that the tramway stopped its services right opposite his house both for him to board and alight from the tram in each direction. The company had agreed a signalling arrangement with him, so that if a flag was raised on a flagpole at the front of his garden near the tramway, the tram would stop to pick him up. If he wished to get off the tram outside his house, he merely needed to inform the conductor when he boarded, who would in turn ask the driver to halt at the Clark house on the main road. Similar arrangements remained in force throughout the existence of the tramway, its 'stations' seeming to be whereever local people wanted them to be.

In July, the tramway and the post office finally reached an agreement for the carriage of the mails. The Postmaster General and the directors of the Wantage Tramway agreed a mutually acceptable charge for the mails to be carried by the tramway. A contract was signed under which the trams would convey the post sacks from Wantage Road for £52 2s 10d per year.

A number of improvements were made on the tramway during the autumn. The Hughes Tram, which had been completely mechanically rebuilt at Swindon the previous year, received a new cab and bodywork at Wantage. The replacement of the engine's bodywork was undertaken by Robinson and Auden, and Whitings, after which the

The Hughes steam tram engine and one of the tramway's passenger cars outside the house belonging to Mr. Clark, the owner of the flour mill. The tram stopped here if a flag was flying from a flag pole which is out of sight signalling that a member of the Clark household wished to board the tram. *Vale and Downland Museum, Wantage*

Hughes engine was in all respects a completely new locomotive which just happened to look identical to how it had prior to its major refurbishment. Another improvement involved the tramway company's engineer arranging to have steps fitted to the platforms of the large tramcar to allow passengers to board from road level, which was necessary given the company's flexible view of where their station stops were.

Despite the signing of the contract with the post office, the tramway saw its income decline during the winter of 1896. The company found it necessary to ask the general manager to reduce the wages of the labourers working on repairs to the line during the winter, and the wages of driver Webb, who had replaced Hewitt as passenger driver, and Spicer, the goods driver, by 1s a week in each case. No mention is made of any adjustment to the wages of the other goods driver, Robert Greenacre, who may either have had no adjustment made since he was paid less than Spicer, or had left the company. No further mention of Greenacre was made by the company so it seems more likely that he was no longer working there.

The end of September saw the company involved in an accident, although on this occasion no blame could possibly have been placed upon the tramway. Whilst in the yard at Wantage Road, the Matthews Tram was damaged when a train of wagons being shunted by a GWR locomotive collided with the stationary tram engine. No. 6 was damaged in the collision, but the Great Western immediately informed the tramway that they were solely responsible for the accident and would be pleased to take the tram to Swindon where it would be repaired at the railway company's cost. In early October, the company having already accepted the offer, and the engine sent off for repairs, asked whether the GWR would be able to turn up the wheels of the tram whilst it was there, at the cost of the tramway. Swindon agreed to do so while the engine was in their care.

Swindon had meanwhile completed the repairs to No. 5, and after the engine had been returned to the tramway during October 1896, sent the bill for the necessary work to Wantage. The locomotive was in a worse state than had been estimated, the GWR's estimate for £262 proving to be well short of the final cost of £327 which the company paid on 5 November.

The spring of 1897 brought with it a need for the tramway to perform urgent repairs to the track on the line, possibly due to the lack to work performed during the winter of late 1896 when the tramway's income had been lower than usual. Traffic had recovered in the early part of the year, and with more funds at its disposal, the company authorised the work on the line.

Finances were still not as healthy as the company directors would like, with many customers' accounts overdue. If the debts owed to the company could be reduced, the financial position of the company would be much improved. The general manager was, to that end, instructed on 1 April to compile a list of overdue accounts which he had been unable to have the customer pay. The list was to be submitted to the directors for their perusal for following week.

On 8 April, the manager's list of difficult to resolve debts was presented to the board, who instructed him to chase for payment of all overdue accounts, and to report to the directors at their 3 June meeting those customers who still refused to pay the monies owed to the company. In the meantime, the general and goods managers were given permission to refuse to deliver any parcels to those who had an overdue account with the company until such time as the outstanding debt had been settled. A large amount of customer debt remained unpaid at the time of the June board meeting, and the manager requested permission to make one final application for payment of the overdue amounts, followed if necessary by referring the debts for prosecution at the County Court. Permission was given, and final demands for payment were sent to those who still had not settled their overdue carriage accounts. The threat of legal action seems to have focused the minds of the majority of the debtors, who promptly paid, although the company found it necessary to take a small group of debtors to the courts during the autumn.

The tramway was involved, partly, in a dispute which arose between Robinson and Auden's Vale of the White Horse Iron Works, formerly Gibbons and Gibbons Wantage Ironworks, and the Wilts and Berks Canal Company in 1897. In November, the road bridge over the canal next to the tramway's canal bridge at Grove had been declared unsafe for heavy road traffic. The canal company was responsible for the

No. 5 crossing the Grove canal bridge with a goods train. The last wagon is a fuel tank destined for the Anglo-American Oil Company's depot at Wantage. Noticeable is the lack of a brake van which short trains were allowed to run without. The road next to the tramway was also crossing a bridge, which was the cause of the disagreement between the canal company and the iron foundry. *John Scott-Morgan Collection*

maintenance of the bridge, with the local road authority only maintaining the road surface on the top of the bridge. The dispute arose because Robinson and Auden used the road to transport their products from their works in Wantage to Wantage Road station using their own traction engine.

The road bridge was no longer able to carry the weight of Robinson and Auden's traction engine and so the company had needed to despatch all of their products to Wantage Road using the tramway, which they rarely did under normal conditions. The owners of the iron works wrote to the canal company in late October to seek compensation for the additional costs they were having to pay.

The canal company's company secretary, W. Ainsworth, had replied on 3 November denying any liability on the part of the Wilts and Berks Canal Company. The letter received a reply on 11 November with the iron works' owners reiterating their position. Robinson and Auden, as licenced users of the road with their traction engine, reserved the right of action against the canal company. The iron works held the canal company responsible for the damage caused by the increased costs of transport using the tramway to bring goods to and from Wantage Road. The closure of the road to them, they said, had been caused by the canal company's announcement that their Grove Road bridge was unsafe. The engineers at Robinson and Auden had examined the bridge and found that repairs had been 'scandalously neglected' and the thin layer of the metalling on the road surface was 'absurd'.

Ainsworth replied on 15 November stating that the matter had been referred to the canal company's directors, but that he had received legal advice from the canal company's solicitors that Robinson and Auden were liable for the damage to the bridge which had resulted in its closure to heavy traffic. He also replied that Robinson and Auden's road licence for their traction engine did not cover the use of bridges not intended for such a heavy road engine. The canal company took the position that all the canal bridges had not been intended, and are not fit, for such heavy traffic as the iron works' traction engine and trailers. The tramway company's bridge clearly had been designed for heavy traffic but the canal company made no mention of it, avoiding mention of the tramway in all of the correspondence relating to the claim.

Robinson and Auden responded to Ainsworth's missive on 3 December. They continued to argue that the canal company was liable for the increased costs. The iron works, they said, had accumulated an increased amount for traffic between their works and Wantage Road each month through the necessity of using the tramway instead of their own traction engine. The company intended to bill the canal company for the excess costs each month until they had announced that the road bridge over the canal at Grove was safe for their traction engine to cross as it had 'for the last fourteen years past'. They also pointed out that since Robinson and Auden's last letter, the road surface had been repaired with a thick layer of tarmacadam and that the company still maintained that the bridge would have still been safe for their traction engine to use 'had not the repairs and road metalling been scandalously neglected'.

The company's solicitors, Llewellyn Jotcham and Son, had already written to the canal company with a legal reply to Ainsworth's letter, restating the liability was with the canal company and requesting again reimbursement of the iron works' additional expenses. Llewellyn Jotcham was also a shareholder and a director of the tramway company; it can only be wondered what he thought about the dispute. Edward Ormond, company secretary of the tramway, was also a shareholder of the canal company. Ormond was no doubt interested in the dispute between the canal company and the iron works since the part closure of the bridge had resulted in a considerable amount of additional goods traffic being placed with the tramway by the iron works.

The canal company responded to the two letters in a response dated 4 December 1897. Ainsworth stated that the canal company still accepted no responsibility for the state of the bridge. Their position was that the roadway was the responsibility of the district authority who had recently, at the time of Ainsworth's letter, resurfaced the road with a new layer of tarmacadam. Ainsworth did not agree with Robinson and Auden's suggestion that the canal company had not maintained the bridge correctly, stating that the bridge had received general repairs needed and there had been no neglect whatever in the case of the road bridge. Ainsworth also stated that his response was the result of legal advice, but that the case had been sent to London for further opinion. Ainsworth finished by informing the iron works that their traffic could not be classified as the 'normal traffic' which the bridge had been built to carry. Meanwhile, the iron works continued to use the services of the tramway company, their traction engine remaining out of steam and it seems that they continued to do so for some time. The dispute between Robinson and Auden and the Wilts and Berks Canal was of benefit to the tramway company, not only through the increased revenue that the iron works' traffic had brought to the company, but also the damage to the reputation of a competitor for goods traffic from Wantage.

The tramway benefitted from their transport rival's predicament, with traffic formerly carried on the canal or using the road from the town to Wantage Road station moving onto the tramway. The canal was still being used for the carriage of freight such as coal where time was not so critical, although the tramway was progressing moving more of this traffic since long distance goods were being placed with the railway companies more and more as the nineteenth century progressed. The declining traffic on the Wilts and Berks canal may have been the reason for the condition of the canal bridge at Grove as the company sought to reduce costs through less regular maintenance among other economies. The tramway at the same time continued to go from strength to strength.

Whilst the company was watching the canal's difficulties develop, issues with the tramway's permanent way had come to fore once again in November 1897. The GWR wrote to the tramway informing them that their engineers, whilst undertaking work in the yard at Wantage Road had seen that one of the tramway's sidings there was in a dangerous condition and in need of urgent attention. The tramway responded that they would take care of the work forthwith, although it was only in January 1898 that the tramway selected contractors to undertake the work.

The year ended with the company opting to have the tramway weeded and arranged for the work to be undertaken by local people, to put some extra money into their pockets. The line was split into short stretches of one chain in length. Anyone who wished to weed a stretch of the line was to be paid 1s 8d per chain for one side of the line, or 2s 6d if they weeded both sides of the tramway. A number of Wantage people signed up to weed the line and earn a little extra cash for Christmas.

1898 started with little to bother the directors or managers, with the operation of the tramway's services continuing without incident. The company was then presented with a new problem with the line in April. The condition of the retaining walls which supported the small bridge over the Pill Ditch at Wantage was causing concern. If the retaining walls were to collapse, which seemed likely, the tramway would have to close the line at that point and resort to road transport. Estimates were sought for the urgent repairs, with the lowest bid from John Gregory and Brothers of 10 guineas accepted. Work began straight away and the danger was averted.

The late spring had presented the tramway with another problem to resolve. One of its drivers, Webb, had been off work sick for several weeks, and by early May had been off work for so long that he was receiving half pay while absent. Having heard nothing from him the company informed him on Thursday 5 May that all pay would be stopped, as they were entitled to do, after 21 May. Nothing more seems to have been heard from Webb, so it can only be assumed that he was dismissed from the company's service due to his lack of attendance at work, or that he had passed away as a result of his illness.

Staffing challenges continued to plague the company. A complaint was received from Gibbs, the GWR's station master at Wantage Road, about the conduct of one of their passenger conductors, William Savory, in late May. Savory had made rude comments to Gibbs and refused to load luggage onto the tramcars. Savory apologised for his behaviour, which Gibbs accepted, although the tramway directors still summoned him to their 2 June meeting to warn him about his future conduct. Savory's conduct was not the only incident to concern the company during May.

A claim was made against the GWR and the tramway relating to the loss of part of a consignment of sugar, with the loss valued at £3 3s. The loss appeared to have been caused by theft of part of the shipment at some point in its journey. An investigation was undertaken by the GWR on behalf of both the railway and tramway, finding that the loss appeared to have occurred after the van had been shunted onto the tramway. No culprit was identified, although all of the tramway's employees were ruled out as suspects. The tramway offered to pay the full amount to the owner, even though the theft appeared to have been committed by an outside party.

Debts became a problem to the tramway company again during the summer of 1898. Watson, the general manager, compiled a list of all the outstanding carriage accounts, and found that the company was owed a total of £352 15s for unpaid carriage bills. Watson spent two months chasing for payment of the outstanding monies, reporting to the board on 15 September that he had managed to recover most of the overdue amount, with £250 paid up. The balance of the debt was to put to the company's solicitor, Edward Ormond, for legal prosecution.

The cause of much of the overdue debt and refusal to pay carriage charges had been the lack of proof of delivery available to the tramway company. Fred Ryman, the goods agent, had once again stopped obtaining signatures for goods in the delivery book which made it more difficult for the company to collect payments. The directors summoned Ryman to their 22 September meeting and reminded him that it was important that he do as he been instructed by the company's goods manager, and only deliver goods if he was able to obtain a signature from the consignee or a representative. If he was unable to have someone sign the delivery book, he was to return the consignment to the goods shed at Wantage pending collection and payment of the charges by the recipient.

1899 was a year of upheaval on the tramway, the end of the century proving difficult. The company refused the application by Charles Clement, who was living in the company's house in Mill Street, to renew his lease at a reduction of £2 per year. The directors asked Watson, the general manager, to live in the house as soon as repairs to the building had been completed. He refused due to his wife's health and the confinement of the premises. The gardens, stables, and outbuildings which accompanied the house were not part of the lease, all of them remaining in use by the tramway, and with no outside space, the general manager had to refuse to relocate his home to the tramway's house. The tramway, with no other tenant lined up, allowed Clement to remain in occupancy at the reduced rent he had requested, a new lease being drawn up for a rent of £14 per year. It was as well that Watson did not take up residence of the tramways house.

In March the general manager had taken a leave of absence from the tramway to look after his wife. The manager's absence from the tramway would prove controversial despite it being authorised by the company directors. A complaint was raised by a shareholder at the annual general meeting for the previous financial year, which had been delayed until 6 April. The complaint resulted in a motion being put to the meeting that the manager had exhibited a lack of diligence, since he had been seen at a race meeting during his leave of absence, and the shareholder concerned had taken offence at him being there during his leave.

There was no reason why Watson shouldn't have attended the race meeting whilst looking after his wife, who may also have been there, regardless of which the motion was passed. Watson had worked hard for the tramway and made many improvements, and was generally viewed as a good manager, but the vote put the directors in a difficult position. The board had no option but to carry out the wishes of the shareholders and ask Watson to resign, or accept notice of termination of his employment by the tramway. Watson opted to resign, which the directors regretfully accepted. The post was advertised, and by the end of the month the company had received a large number of applications, from which six men had been selected to come to Wantage for an interview on 15 May. The invites were sent out but only four men replied to accept.

The four men who came to Wantage to interview for the vacant general manager's position were a Mr. Blake from the Rhondda Valley, whose third class return train fare was paid for by the company; A.J. Haslam from Reading, who received no expenses; G. Cook from Darwen in Lancashire who received £2 toward his expenses; and a Mr. Stapleton from Bournemouth whose third class train fare was paid.

All four were interviewed by the directors, with William Noble from Wantage also seen. The directors concluded that Cook and Stapleton were the most likely two candidates. Cook had the best qualifications, but the directors delayed making him an offer pending further enquiries about him with his referees. The replies that were received from Cook's referees and the further enquiries taken to establish his qualifications raised doubt, and the directors decided to interview a sixth man, Mansfield Collins, who had come to their attention since the 4 May interviews. Collins met with the directors, who were impressed with his experience and plans for the tramway. His testimonials were extremely good, and the tramway offered Collins the position, which he accepted.

Mansfield Collins's tenure as manager of the Wantage Tramway did not last long. Collins informed the directors, less than two weeks after he had accepted the post, that he no longer wished to continue to work for the company. His previous position was still open and he wished to return to it, and wanted to leave the employ of the Wantage Tramway on 31 May. The directors sent a telegram to Stapleton, who they had interviewed at the beginning of the month asking if he was still interested in taking up the position at Wantage. Stapleton replied that he was, and as he was also 'disengaged' at the time, he would be able to take up the role at the tramway company's convenience. Stapleton was asked to come to Wantage the following Monday, 29 May as the manager on a one month trial, which he duly did.

The tramway had still not found its new manager, Stapleton informing the directors on Tuesday, after just one day in the position that the duties was not suited to him, and the work was more than he could undertake. Both Collins and Stapleton departed Wantage on Wednesday 31 May, leaving the tramway managerless once more. The position was advertised again in the Reading Mercury and Oxford Chronicle, the directors deciding that they preferred to try and recruit a local man for the role.

William Noble reapplied, and was deemed the best candidate out of all applicants. He was asked to come to Wantage to meet the directors again which he duly did on 8 June. W.H. Noble was, at the time of his interview, employed in the goods manager's office in Cirencester, on the Midland and South Western Junction Railway. He was offered the position on the Wantage Tramway and agreed to take up the job on 19 June on a salary of £110 per year. The company's management woes were not over.

James Sims, the company's goods manager, attended the 29 June directors' meeting at his own request. He explained to the directors that he was no longer a young man and acknowledged that he was no longer able to undertake his duties as goods manager because of his advanced age. He asked the directors to accept his resignation and hoped that they would assist him in obtaining an alms-house for his retirement. The directors did help Sims in his retirement and arranged for him to be given a house.

Alms-houses were charitable institutions which aimed to provide a home for poor or older people in need of a place to live without having to pay rent. The residents of alms-houses had no security of tenure, being purely dependent upon the goodwill of the trustees. The intervention of the directors of the Wantage Tramway Company were doubtless helpful in him not only being granted a 'retirement' home, but their continued interest secured his residency in his new home for the rest of his life after leaving the company.

The tramway was once again in need of a new manager, but on this occasion had a ready replacement already working for the company. Lovett had been working as Sims's assistant for a few years and was seen as the ideal man to take up the reins in the goods department. He was offered a promotion to manager on a wage of £1 per week, which was accepted.

With their management woes resolved, the directors may have thought that their problems were over, but that was not to be. During the summer, the Manning Wardle locomotive began to need repairs, and could no longer be patched up by the tramway's engineer. The engine was eventually sent away to Swindon for a heavy overhaul. When the locomotive arrived at Swindon, the Great Western's engineers reported back to Wantage that the boiler was in a poor state and would need to be replaced. The tramway wrote to Manning Wardle in Leeds, who replied that they would be able to construct a new boiler for the engine. Discussions continued and it was agreed that the locomotive would be sent on to Manning Wardle who would undertake the repairs to the engine as well as fitting a new boiler. At the end of August, Manning Wardle sent the tramway an estimate of costs. The new boiler and firebox would cost the company £350, with the repairs to the specification received in Leeds from the GWR's engineer estimated to cost the tramway a further £600. William Dean remained involved in the repair work undertaken on No. 7, despite it not being undertaken in Swindon. The standards to which the repairs were being carried out and the costs involved were personally checked by Dean on behalf of the tramway, an arrangement which had been agreed on 31 August at the tramway's board meeting.

In the meantime, the tramway had offered their erstwhile goods manager another six months work as he began his retirement. Sims agreed to stay on in a different role with reduced workload for the wage of £1 a week. One of the arrangements made whilst Sims was still working in the goods department at Wantage saw a change to the first freight service of the day. The company had received frequent requests from businesses in Wantage and the surrounding area for an earlier delivery time in the mornings. Sims and Lovett arranged for the 'goods engine' to bring up the London goods earlier each day, which allowed the goods shed to sort the goods and put them out for delivery earlier. Sims eventually left the tramway for the last time at the end of the six months on 22 January 1901, at which point Lovett's wages were increased to £1 5s a week.

The motive power situation on the tramway deteriorated further as the nineteenth century drew to a close. The Hughes tram engine was again in need of repairs by October, and so William Jackson, one of the tramway company directors travelled to Swindon to meet William Dean to discuss the state of the engine. Dean agreed to carry out repairs on the tramway's engine when they could send it up to Swindon. With two of the tramway's locomotives out of service, No. 7 at Manning Wardle in Leeds and No. 4 to shortly be sent to Swindon, Wilshire examined the Matthews tram engine and No. 5 to check on their condition.

The new general manager, Noble, approached the board of directors in early October, requesting a pay rise of £10. The directors declined Noble's request, informing him that they thought it was premature since he had only been in the role for three months. In retrospect, perhaps this was the first indication of events that would follow in the early twentieth century.

The last three months of the year brought more challenges to the tramway. Complaints were received about the state of the line, with regular travellers, including a Mr. Houghton who complained on 26 October, telling the company's manager that they thought the line was dangerous. Nothing was done to repair the line, which was reaching a bad state. Noble complained about the way that goods were being delivered in the town by Fred Ryman and asked the directors for permission to cancel the agency agreement with him, which was granted. Noble then agreed a new delivery agency contract with William Griffin which began on Monday 12 October.

The company's general manager raised another complaint, this time with the GWR. Noble complained that the tramway's engine was involved in a large amount of work shunting trucks at Wantage Road, before the loaded goods wagons for Wantage could be reached. Issues on the tramway would continue into the new century.

Chapter 9

INTO THE TWENTIETH CENTURY

The new century started much as the previous one had ended. The levels of service on the line continued to deteriorate under Noble's management. Claims for damaged goods or for non-delivery continued to be received by the company, with the directors accepting and paying them. It seems that the directors were becoming concerned with the performance of the company and small changes were made to attempt to improve the situation. One of the changes that the company made was in response to a concern about how the passenger luggage arriving at Wantage was being delivered to the passengers' homes. In order to ensure that the work was being completed correctly, the company purchased two badges, marked W.T.C. No.1 and W.T.C. No.2, which were to be worn on the arms of the two boys employed to carry luggage and parcels.

Despite the decline in the tramway's service levels, the general manager approached the board once more in late January, requesting an increase to his salary. The request was discussed and since the directors were providing rises to other members of staff, Noble received the £10 increase that he had asked for. The goods manager, Lovett, without requesting a salary increase, was awarded a considerable rise, which took his salary from 25s a week to 30s (£78 per year). The passenger conductors also received an increase of 1s per week to their wages.

The company's difficulties continued into the spring, with serious damage occurring to the canal bridge. The damage occurred when a tank wagon owned by the gas works, carrying ammoniacal liquor, derailed as it crossed the bridge, seriously damaging the timbers between the rails and causing lighter damage to two of the bridge girders. It appeared that the tank wagon was in a poor state of repair, Wilshire's report being less than complimentary about the condition of the truck. As a result, Noble was instructed to discuss the incident with the gas works manager, and to inform him that the tramway would no longer be prepared to carry liquor in the tank wagon until the GWR was prepared to accept the wagon onto their line. The tramway was effectively demanding that the gas works tank wagon should be repaired to main line standard, or else they would not take it in their goods trains.

Nos. 5 and 7 outside the old running shed at Wantage. Only No. 7 appears to be in steam. The overgrown condition of the yard makes an interesting comparison with the condition of the tramway's facilities within the GWR's yard at Wantage Road. The cylindrical structure behind the locomotives is part of Wantage Gas Works. *John Scott-Morgan Collection*

The gas works agreed to repair the wagon to the GWR's standards, and once the repairs had been completed the empty truck was taken to Wantage Road for inspection by the GWR's engineers, who certified it fit for use. The Wantage Gas Company had also paid the tramway for the repairs to the canal bridge.

In the late spring, the company found itself once more searching for a new manager, when the goods manager resigned in late May. Lovett asked to leave the company since he had found a better paid post elsewhere which he wished to take up. His resignation was accepted, and Noble was instructed to find a successor. The GWR divisional engineer added to the company's problems at the end of the month when he informed the tramway that one of the sidings in the yard at Wantage Road was in very poor condition, and that they required the tramway company to undertake repairs. If these repairs were not undertaken soon, the GWR would carry out the work and bill the costs to the tramway. It seemed that the tramway was generally in a poor state of repair by 1900. The GWR did send a bill to the company shortly after the engineer's report, although that invoice was for repairs to the goods engine, No. 5, which had been completed at Swindon. The cost of £333 13s 6d was paid from the previous financial years profits of £2,116 4s 9d. The high level of profit may have been due less to an increase in traffic, but the general manager's lack of willingness to authorise repairs to the tramway line.

The company received some relief from its problems in June with the appointment of two new members of staff. Joseph Bullock joined the company as goods manager in the middle of June, accepting the position with a wage of 28s a week. He had been employed on the GWR and received glowing references from stationmasters Bould and Rumbold, then at Marlborough and Swindon respectively who had both worked with him. The company had also been looking to employ a 'gauger' to maintain the track on the tramway, and shortly after Bullock joined to company found a man for the job. Charles Bint joined the company from the GWR, where he had been performing the same role.

The good news then stopped. Griffin gave notice that he wished to discontinue his cartage agreement and give up his goods agency with the tramway. The company decided to undertake deliveries itself and purchased Griffin's horses and cart for £25. At the same time, the company employed a policeman to patrol the private part of the line and prevent trespassing, as this had become increasingly common. The policeman joined the company at the end of October and set about protecting the company's yards.

A group photograph of the Wantage Tramway Company's delivery staff and the company road vehicles at Wantage. The yard behind the group seems to be quite full of railway wagons. *Vale and Downland Museum, Wantage*

As the year drew to a close, the general state of bad luck improved, no doubt with relief from all the staff working on the tramway. The tramway operation returned to a quiet state of efficiency with Joseph Bullock taking over the goods services, which ran much more smoothly, as did the passenger services. Charles Bint made good progress with repairs to the worst parts of the line, although the remainder was by no means in a perfect condition, and more importantly no one resigned from the company's employ. The goods services had improved to such a great extent, and very few complaints for non-delivery or damage to goods had been received, since Bullock had become the tramway's goods manager that the company awarded him a 2s a week pay rise, taking his wages to 30s (£78 per year) after only four months' service.

The new year started with the tramway receiving a bill from the GWR for repairs to the siding which had been in bad condition at Wantage Road. The tramway had not undertaken any repairs, the company engineer being occupied with rectifying damage to the canal bridge at Grove, still open to rail traffic with a temporary speed restriction, which had been caused by the gas works' wagon, completed by 31 January, and the general manager not authorising anyone to do the work. The GWR had completed the repairs as suggested by their District Manager and the costs had been invoiced to the tramway. The invoice from Paddington amounted to £46 13 which the tramway directors delayed payment of pending discussions with the railway about what they viewed as the excessive cost incurred in the renewal of the siding.

The disagreement with the GWR dragged on, with both sides failing to reach any agreement. The railway asked the company again to pay their bill at the end of February, with the tramway directors stating that the charges were unreasonable and also that the GWR continuously used it, whilst the tramway rarely had use of it. The tramway was asked again to pay the bill in September, with the tramway offering to pay half. The railway then followed up with another bill for work on the siding in October for an additional £24 7s for work on the same siding. The railway was informed that the tramway would only pay half of the total charges of £70 2s 2d that had been presented to the company. The dispute was finally resolved at the end of November, when the Great Western's general manager wrote to the tramway accepting their payment of half the bill and offering to write off the remainder. A cheque was duly sent to Paddington.

Whilst the dispute with the GWR was raging, the tramway had been making improvements to its goods facilities. A new delivery cart had been purchased from the Burstal Wagon Company for £16 to add to the company's fleet of delivery vehicles, with a new driver employed to man it. With space in the yard at Wantage becoming cramped with the growth of the line's goods traffic, plans began to be made to expand the tramway. An offer of £5 was made to the Wilts and Berks Canal

No. 5 shunting an LNWR steel sheet wagon at Wantage. The crane on the goods platform behind the wagon was purchased from the Wilts and Berks Canal Company and proved extremely useful to the tramway when handling traffic such as steel sheet. *John Scott-Morgan Collection*

Company for the redundant crane on the canal wharf which was accepted. The second hand crane was then moved to the tramway's yard.

Rapidly outgrowing the yard at Wantage, it was becoming clear to the tramway directors and the goods manager that additional space was needed urgently. A plan was drawn up by James Bullock and presented to the board, who authorised the project. The additional land needed for the expanded goods facilities was found just to the west of the existing yard. The land, owned by John Simmons, was purchased, and an agreement was made with Edward Ormond for lease of a stretch of land adjacent to Grove Street for an extra siding early in 1902.

The tramway appeared to have once more been having difficulties in delivering goods punctually from Wantage Town station during the year, but this time caused by the high level of goods traffic arriving on the tramway. It was reported at the Annual General Meeting of the company on 24 October 1901 that it had been necessary to expand the company's delivery capability since delays were being caused at Wantage Town by the lack of road vehicles to take goods on to their destination soon after they had arrived in the yard. The company had purchased an additional four wheel 'trolley' (cart), and a new horse for a total of £58 in order to improve the delivery capacity of the tramway earlier in the year. Increasing demand for goods deliveries as the company's goods traffic continued to grow meant that the company needed to invest further, acquiring an additional horse and a two wheel cart for lighter deliveries, purchased for a combined amount of £56 13s 9d. In August, the tramway purchased another road cart from Kendall and Wright for the sum of £20 which was used to expand the company's delivery capability which they were still undertaking instead of appointing a delivery agent.

The company had, despite the challenges it had faced with deliveries during the year, had a good year. Passenger traffic receipts had increased by £78 2s 11d to £991 14s 8d, whilst parcels and goods traffic receipts had increased to £173 13s 10d and £1,144 19s 3d respectively. The company had carried 8,729 tons (up by 1,146 tons) of 'truck goods' (heavy freight such as coal and iron), and 1,635 tons (up by 170 tons) of 'trade goods' (food, clothing, household goods). Profits had once more been good, and the company was able to announce a dividend to the shareholders for the financial year which had just ended.

1902 started with the company making some changes to the accesses to the yard at Wantage Town. Following the trouble that the tramway had in the past with 'gangs of boys' gathering in the yard on Sundays when the tramway wasn't running, where they vandalised buildings and interfered with the points, the company had added barbed wire to the fence. To improve the security of the yard when the trams were not operating, on 30 January 1902 the tramway fitted two sets of iron and wood gates on the two entrances to the yard at a cost of £8 10s which had been made by Thomas Whiting. The two entrances that were now gated were both on Grove Street. The first was at the Mill Street end of the road at the end of the first block of buildings and was the main road access to the yard. The other was the point at which the tramway's line to Wantage Road emerged from the yard and crossed Grove Street.

A further improvement was needed on the tramway after a truck in the goods train derailed on the curve at the approach to Grove canal bridge. The curves on the approach to the bridge were still causing trouble, despite the work which had been done to increase the radius of the curves. Derailments of freight stock appeared to be a fairly regular occurrence, and the company, in one last effort to try to improve the situation, fitted checkrails to the curves to prevent rolling stock jumping the rails. The checkrails seem to have resolved the problem with very few derailments reported at this point on the tramway afterwards.

An issue developed among the tramway's management in the late spring when Wilshire went off sick, prior to which the directors had begun to have concerns about the quality of his work. Wilshire had reported his illness on 24 April but by 29 May had still not returned or contacted the tramway. The directors therefore summoned him to attend the next directors' meeting to be held on 3 July. He duly did so and was informed that the directors were not satisfied with his work and were concerned about his long illness and wished to know if he was recovering.

Grove canal bridge, November 1930. The checkrail installed in 1902 can be seen in the foreground near the left rail. *John Scott-Morgan Collection*

Wilshire seems to have announced that he was physically fit for work and admitted that he had been for a short while but had family concerns to attend to. He did, however, accept the directors' comments about the quality of his work which had been caused by a variety of factors.

It was agreed to allow him a leave of absence in order that he could resolve his family issues before returning to work. The company and Wilshire agreed that, to give him more flexibility, he would no longer be paid a salary, but instead would receive an hourly wage of 8d which would be paid against timesheets that he submitted to the manager each week. Wilshire was also capable of earning more for the time he spent working at the tramway, which under his previous arrangement had not been paid. The new arrangement also allowed him to leave the tramway for a short period of time to return home if he was needed. The new agreement seemed to have suited everyone.

In early July, the company's rolling stock received a cosmetic upgrade, when Tom Richens was given a contract to paint, line out and varnish the two passenger engines as a cost of £4 18s. The two engines mentioned in the invoice were most likely the Hughes and Matthews steam trams. The two locomotives were both needed for the passenger traffic, which was still growing, whilst No. 5 and No. 7 were busy on the goods services. The increasing levels of traffic, especially goods, created some challenges.

At the end of the financial year, the general manager was awarded a bonus of £10, but Noble was still unsatisfied with his earnings. He approached the directors requesting a pay rise to take his salary to £140 per year, but the board offered an increase to £130 with a bonus to be paid each year if profits allowed.

The tramway was presented with an unexpected cost to deal with in late December. The Manning Wardle goods engine, No. 7, was found to have worn the tyres on its wheels down to a level which could not be returned, and the GWR was approached to fit new tyres to the locomotive.

During 1902, the tramway's traffic figures improved again. The financial year ended with the tramway having carried 43,678 passenger journeys, 1,917 tons of trade goods and 11,390 tons of heavy freight, an increase over the traffic carried in 1901. The profits for the trading year were very good and the company was once again able to pay a dividend to all of its shareholders for the 1900-1901 accounting year.

Locomotive trouble plagued the tramway as soon as the new year began with the Hughes engine needing a boiler retube as a result of a visit by the company's boiler inspector. The company purchased eighty new tubes for the locomotive in late January and engaged the services of a local boiler smith to perform the repairs on site in the yard at Wantage.

Despite the fault with one of their passenger engines, the tramway's management felt confident in being able to expand the passenger service. The tramway advertised an additional service to meet the 11.32 Up train every day, instead of just on three days of the week, to meet demand from the large volume of passengers now using the late morning train. The GWR later planned to change their timetable, adding in an additional stop for the late morning Down train, informing the tramway that from 1 July 1903 it would stop at Wantage Road at 11.21. The tramway company changed their timetable, sending the tram to meet the new service with the tram leaving Wantage Town at 11am. The tram service would also then wait for the arrival of the 11.34 Up train before departing for Wantage Town. The company made arrangements with the GWR for a further change in the summer. The Great Western had continued to run a number of special excursion trains from London and Bristol, among other places, and for the summer of 1903 were planning a series of excursions to Wantage, which would of course terminate at Wantage Road. The tramway company agreed to offer a special return fare of 9d for persons travelling to Wantage on the GWR's special trains. The additional revenue from the full trains of day trippers boosted the tramway's passenger income for the year considerably.

The early autumn arrived, as did the boiler inspector. The inspector had come to Wantage to check the boilers on the two freight engines, with those on both No. 5 and No. 7 given a clear report, although the inspector mentioned a small number of minor repairs which Wilshire was already aware of and had planned. The inspector passed both the locomotive boilers for a further twelve months but recommended that the slight repairs should be done as soon as possible. The tramway's engineer was instructed to complete them by the directors after the 16 August board meeting.

The company's passenger tram cars were beginning to age and were in need of either expensive repairs, or replacement. The chairman of the tramway attended a sale of horses and tram cars that Reading Corporation Tramways were running to dispose of old redundant cars and horses which were no longer required. Wilshire also attended the sale in order to inspect the tram cars which were being sold before the tramway placed any bids. The company's representatives purchased two of Reading's redundant tram cars for £11 11s and £16 5s 6d. The 'new' tram cars were taken to Wantage by the GWR on wagons, since Reading's tramway had been built to a narrow gauge. The Wantage Tramway Company arranged for the GWR to send a fitter to Wantage to take off the tram car's wheels and regauge the trams to run on their tramway. New wheel sets were on order from a local foundry which would be fitted by Wilshire.

The Matthews tram engine with passenger car No. 5. The passenger car was one of the two purchased from Reading Tramways in 1903 when Reading abandoned using horse powered trams. Originally a two deck car, the upper deck was removed and the seats sold. The car has been regauged, and of note are the new wheels and springs supplied and fitted by the GWR. *John Scott-Morgan Collection*

The wheel sets and axle arrangement on the Reading cars were later found to be unsuitable for running on the Wantage Tramway, and the GWR was once more approached for assistance in September. Swindon offered to rebuild the tram cars with new wheels, axle boxes and springs to make them more suitable, which the tramway accepted. The new parts were made and sent to Wantage where they were put onto to the cars by GWR fitters. The rework of the two Reading cars cost the tramway £35.

In 1903, some of the company's traffic had grown once more over the year. Passenger figures were up to 45,438 journeys, whilst trade goods tonnage had increased to 1,996 tons, although heavy freight tonnage had decreased to 10,198 tons. Despite the decrease in heavy freight traffic during the 1903 financial year profits remained strong, allowing the directors to declare a dividend.

The beginning of the new financial year saw the tramway's general manager contacting the company directors for yet another pay increase, which was discussed in their 27 November 1903 board meeting. He wrote that he was not happy with his salary and requested that it be raised from £130 to £160. Noble's request was only partly granted with his salary increased to £150, but with a bonus of £15 to be paid for the past financial year in his December salary. At the same directors' meeting, the board agreed that the station house at Wantage needed improvement, and that firms of local architects should be approached in order to obtain plans for a new building at the company's headquarters. Work on the designs for the new offices and station house continued into the new year.

Plans for the new station building at Wantage had been drawn up by the beginning of 1904 and the company put out a tender for the construction of the new offices. Three acceptable bids were received from local firms of builders. Whiting and Gregory bid £323 whilst Wheeler and Company offered to build the new station offices for £321 18s 9d. The winning bid was received from Jacob Padburn, who offered to complete the erection of the tramway's new headquarters for just £240, which was accepted subject to the approval of Grant, the tramway's retained architect. Whilst the bidding process had been progressing, Grant had met with one of the tramway's directors, William Jackson, at the architect's practice offices in Reading, to explain the differences in cost between the architect's estimates and the tenders that the company had received. The bid from Padburn was, according to Grant, about right and in the range of cost he expected, with the architect informing Jackson that he also believed the bids from the other contractors were excessive. The contract was accordingly placed with Padburn on 12 January 1904 and construction of the Wantage Tramway Company's new headquarters began shortly afterwards.

The owner of Clark's flour mill at the bottom of Mill Street in Wantage approached the tramway company requesting that a branch be made from the tramway into his mill for the carriage of materials in and finished milled flour out of the works.

The tramway company agreed in principle to construct a branch for the company, provided that they were prepared to guarantee to place 600 tons of freight with the tramway each month for a period of ten years. In return, the company would build the line at their own cost, and offered to carry Clark's freight over the line to Wantage Road for 10½d per ton. The agreement between the two companies was made, with one additional clause, which was never invoked, under which Clark's were to pay the tramway the sum of £400 at the end of the ten-year term, if the miller was to discontinue using the branch after the term of the contract.

The tramway and the miller met at the company's offices to discuss the project further. It was thought likely that local people, and possibly the council, would object to the tramway placing rails across Mill Street, both parties agreeing to seek legal counsel, sharing the costs equally, in order to establish the legal precedent for gaining permanent right of way across the road which would allow the laying of rails into the flour mill. Once the legal situation had been established, the new branch into the mill could potentially be built, but in the short term the project was put on hold awaiting legal advice.

The tramway's line was potentially impacted by a decision taken by the GWR in early 1904. The railway company had decided that increasing traffic on their main line required a rebuild of Wantage Road station. The GWR district engineer's office had contacted the tramway in early February to inform them of the Great Western's plans and invited the company to a meeting to discuss the implications. Two engineers and an official from the GWR were meanwhile sent to Wantage Road to examine the station area with a view to reporting the accommodation that could be offered to the tramway when they built the new station. In the end, the Great Western's final plans for the new station had no impact on the tramway whatsoever, no doubt to the relief of the tramway's management.

Meanwhile, the tramway had increased the frequency of its morning trains. On 25 February it was decided to add two additional departures from Wantage Town to meet GWR trains at Wantage Road. The first, scheduled to leave Wantage at 9.15am, was to meet the GWR's 9.45am Down train, with a second departure to leave Wantage shortly after the first, at 9.35am to meet the GWR's Up train which arrived at 9.59am. The improved service was advertised to start from 1 April 1904.

The construction of the tramway's new headquarters building in Mill Street in Wantage, which still stands in 2024, had been completed by the summer, but shortly afterwards the contractor ceased trading. The company owned by Jacob Padburn went into liquidation with a high level of debt. The tramway was summoned to attend court in London on 5 July to pay the debt due to the erstwhile company to the court so some of it could be used by the liquidators to part pay the builder's creditors. The liquidators agreed a percentage of the debt to be paid by the tramway, a payment of just £8 1s 3d to be paid for the finished building. A cheque was duly sent to the court in London on 8 August, and the rest of the debt was written off. The tramway had got its new headquarters for a bargain price.

The boiler inspector visited Wantage in early August to undertake the annual inspections on the tramway's motive power fleet. The report certified that the boilers on No. 4, the Hughes tram engine, No. 5, 'Jane' and No. 6, the Matthews tram engine, were all in good condition although some minor repairs were needed, which Wilshire undertook in short order. No. 7's boiler was not inspected on this occasion as it still had plenty of time left on its ticket.

The 1904 financial year saw heavy freight traffic return to the tramway with strong growth experienced over the course of the year ended 30 September 1904. The company accounts showed that a total of 16,625 tons of heavy freight had been carried over the line over the past twelve months. Passenger traffic continued to increase with the tramway taking fares for 50,627 journeys along the tramway, whilst trade goods tonnage grew once again to a total of 2,270 tons.

Profits had once more been good, and the tramway was again able to declare a dividend for its shareholders. Another year had passed in which investment in the tramway had yielded, as it had for nearly every year since its inception, a good return on investment.

The year ended with the tramway returning to the project to link Clark's mill with their line. Legal advice had been received and plans drawn up for where the tramway's branch should be laid. In order to begin construction, the land on which it was intended to lay the rails was needed, and so Edward Ormond was authorised on 5 December to approach the owner of the orchard located adjacent to the canal wharf at the lower end of Mill Street with a view to purchasing the entire parcel of land. Mrs. Langley was to be offered £900 for her orchard on condition that Clark would sign the agreement suggested at the beginning of the year. The contract was signed, and Ormond proceeded to agree purchase terms with Langley for the orchard, which was at that time occupied by a third party on an annual lease.

The tramway amended its passenger timetable once again in late December. The GWR had changed its timetable, with the 6.49pm Up train retimed to arrive at Wantage Road at 6.39pm. The tramway added an additional service to their timetable with a tram being sent to the GWR's station to meet the retimed main line train, scheduled to leave Wantage at 6.15pm.

An accident had occurred on the tramway on 5 November which the company had been investigating. Mr. Hudson's son had accidently fallen off the tram car on the journey along the line and had been killed. The tramway was not at fault for the accident, which Hudson had maintained throughout the investigation. The company directors decided at their last meeting of 1904 on 5 December that the tramway would pay for the funeral of the boy and settled the funeral director's £6 bill on the family's behalf.

Negotiations to purchase the orchard from Langley were still ongoing into the new year, with the initial offer of £900 refused by the owner. Ormond was given permission to raise the offer in early January 1905. At the same time, the general manager was instructed to contact the Wantage Engineering Company and other firms who were regularly transporting coal for their own use along the tramway. The manager was to offer all these companies a discounted carriage rate of 1s per ton rather than the advertised 1s 2d charge if they were sending 200 tons or more along the line each month. Other financial changes were made early in 1905. Spicer, one of the company's drivers, requested and received an increase of 1s per week on his wages. The company also entered into a new agreement with Weedon Brothers for the continued lease of the coal yard at Wantage for a further seven years at a rent of £35 per year.

Negotiations for the orchard owned by Langley finally came to fruition at the end of March, with the owner accepting £1,000 for her land. The transaction was quickly completed. The company, not needing the entire parcel of land, offered the tenant a lease on the majority of the orchard for £20 per year, but with the first year discounted to £15 in return for granting the tramway company immediate possession of the land. The tenant agreed since he would only lose a small passage through the orchard on which the tramway's branch was to be built, and retained access to both parts of the orchard either side of the line. The tramway immediately began to construct the flour mill branch. The new line diverged from the existing tramway at the point where the line entered the existing yard, crossed an unnamed narrow pathway and turned to run parallel to the canal along a route through the orchard which is now Limborough Road. The tramway branch then entered the new bottom yard in which a short run round loop was provided along with two sidings, the shorter of which terminated at an L-shaped office and shed. The second siding continued further to the edge of the orchard and created a mystery.

The siding lined up with the short road which entered the yard from Mill Street, aligned perfectly on the 1910 ordnance survey map to continue down the lane and across Mill Street and into the premises of Clark's Mill. The line opened on 1 July and according to the minutes of the 3 July director's meeting traffic had started not from 'the lower yard' but 'between Clark's Mill and Wantage Road'. The comments made in the board meeting seem to suggest that the tramway had put rails across the road and into Clark's mill, although no photographic evidence seems to exist of the level crossing. The 1910 Ordnance Survey map for Wantage does not show the siding continuing out of the yard but the previous Ordnance Survey map from 1900 does not show the Wantage Tramway at all, even though it had been open and operating for twenty-four years. The mystery of whether the crossing was built must remain unsolved. Although the evidence from the company archives suggests it was built, no other corroborating evidence has been found, perhaps the crossing

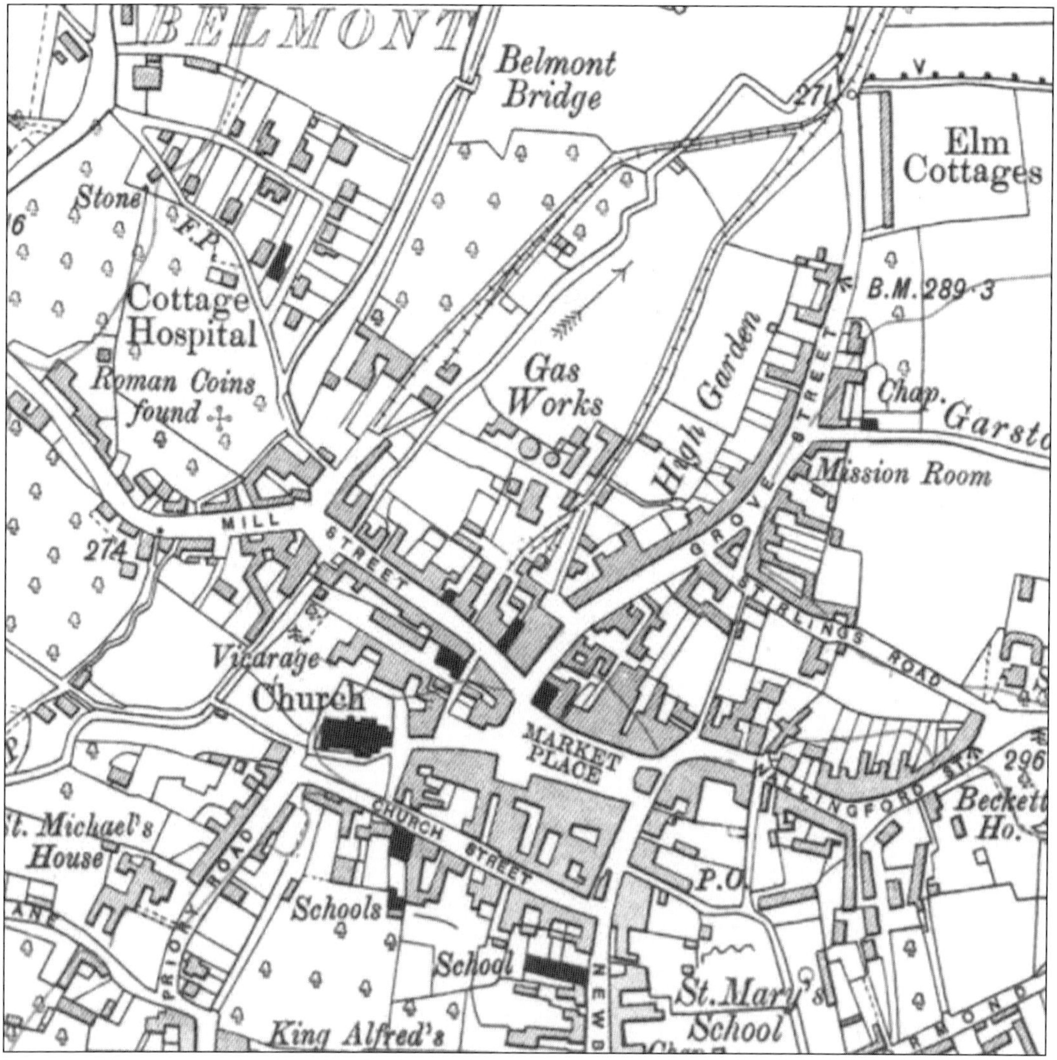

Wantage town centre with the tramway's new yard branching from the line at the top right of the map, and the new yard to the left of the original terminus yard. *Excerpt of Ordnance Survey Six-Inch Series, Berkshire Sheet XIV.SW, 1913. Reproduced with the permission of the National Library of Scotland*

was built but only as a temporary measure to assess its impact upon Mill Street and was later abandoned.

The new yard was, however, definitely equipped with a crane. The company, as had been the way since the very beginning, purchased a second-hand piece of equipment. An offer of £20 was made for a crane located on the canal wharf which was owned by the Wantage Brewery Company but no longer used. The bid was accepted, and the crane was moved into the mill branch yard where it was erected by

A general view of the lower yard at Wantage with a number of coal wagons on the siding to the left. The roadway led out onto Mill Street emerging opposite Clarks Flour Mill. *John Scott-Morgan Collection*

A portrait of No. 7 at Wantage. The locomotive's short wheelbase, ideal for the tramway can clearly be seen. The Manning Wardle engine now has its large cab, added during a visit to Swindon for overhaul. *John Scott-Morgan Collection*

the Wantage Engineering Company at the beginning of September. The engineering company soon had another job to do for the tramway.

No. 7's boiler ticket was coming to an end during October and with the boiler inspector finding that the boiler needed a repair, the company decided that the engine should have repairs done locally by Wantage Engineering. A patch was put on the boiler and the locomotive remained in service on the tramway, but by early December the boiler needed more repairs. The 'new' crane in the lower yard was used to lift the boiler off the engine and it was placed on a wagon and sent away to Swindon for repairs.

The financial year ending on 30 September 1905 saw the tramway's traffic continuing to grow. Passenger numbers were once more up on the previous year at 54,976 journeys, whilst trade goods traffic had grown to 2,378 tons over the year. Heavy freight had grown, but less dramatically than in the previous year, with the tramway's goods trains carrying 17,727 tons of goods. A good profit was returned for the year which allowed for dividend to be declared by the directors.

1906 began with little incident, traffic being good, and no incidents of any note being reported to the management of the company until early March when No. 7 was damaged. Whilst hauling a goods train, it appears that water carried over from the boiler into the cylinders, with one being damaged by the hydraulic action on the piston and cylinder. The damage required a major repair and the locomotive had to be sent to Swindon. The loss of No. 7 from traffic left the tramway a little light on motive power, and the GWR was approached to ascertain if they would be able to loan the company a locomotive whilst the Manning Wardle engine was in the works. The GWR had a suitable engine which they were willing to hire to the tramway company for a fee of 3s per working hour, or £1 per day minimum. The tramway company accepted the offer and No. 1350 arrived at Wantage to cover until No. 7 returned. The Manning Wardle returned to the tramway in late June accompanied by a bill for the repairs amounting to £138 12s 9d.

The general manager of the tramway wrote to the directors on 29 October 1906 to request yet another increase in his salary. Noble stated that he had been with the company for between seven and eight years, and during that time he believed that he had made many improvements to the operation of the company. In return for the work that he had already done on behalf of the company, he requested that his employment be covered by a five year agreement, starting on 1 January 1907, to replace the annually renewed contract that was currently in place. Noble requested that the new agreement should start with an increased in salary of £180 per year and a second increase on 1 January 1909 to £200 per year. He wrote that he had asked for the increase in his pay as work on the tramway had increased dramatically and he had needed to devote a large amount of his own time to the company's business. He also stated that there was a large call for tramway managers with much less

No. 7 at Swindon Works awaiting overhaul. *John Scott-Morgan Collection*

work and very much better wages; despite his veiled threat the directors refused his application for a raise.

The financial year ended well once more with the tramway reporting continued growth across all of its traffic. Passenger numbers were still increasing with 56,180 journeys completed in the year, whilst goods traffic was also continuing to show growth. Trade goods accounted for 2,506 tons and heavy goods for 20,700 tons of freight carried over the last twelve months. Profits were once more very good, and the directors were able to declare a dividend.

The tramway company unfortunately stopped reporting passenger figures after the 1906 Annual General Meeting, and so traffic volumes for the passenger side of the business are hard to find. The report of the directors for the shareholders did however still refer to a summary statistic for 1907 and compared passenger traffic to the previous year. It seems that the decline in the tramway's passenger traffic began before the First World War, the company reporting that its passenger traffic was down on the previous year with fewer journeys having been completed on the trams. Goods traffic remained strong, however, with 2,539 tons of trade and 21,611 tons of heavy goods being carried over the line during the previous financial year. The freight traffic on the line had once more continued to grow despite the small contraction reported in the tramway's passenger numbers.

The financial year to the end of September 1908 once more saw the directors reporting a decline in passenger traffic, although the company annual report blamed the loss of passengers on the inclement weather which had been a feature of a particularly wet summer. Goods traffic also declined slightly during 1908, with the tramway carrying 2,405 tons of trade goods, 134 tons less than in 1907. Heavy freight declined similarly with 20,760 tons being carried over the year, 851 tons less than the previous year.

The tramway had mixed results for its 1909 financial year. The number of passenger journeys were again reported as down on the previous year, with the added drain on the passenger income of the repairs that had been needed to the passenger engine earlier in 1909, costing £300. The decline in passenger traffic was once more blamed upon the inclement summer weather experienced that year. Goods traffic, however, had bounced back incredibly strongly. Trade goods traffic had grown to 2,615 tons, and heavy freight to 26,310 tons.

A worrying trend was developing for the tramway at the end of its 1910 financial year. The annual report showed that the passenger statistics had declined for the third year in succession and despite the directors suggesting that the passengers would return, and that the decline was due to yet another wet summer, it is clear that local people were moving away from using the tramway to travel between Wantage and Wantage Road station. The goods traffic however had held up well over the year being reported as having increased, although the directors had decided not to state tonnages in their annual report from 1910 onwards. The company's finances had been drained by the need to undertake urgent repairs on two locomotives during the year, amounting to a total cost of £300.

The annual report for 1911 showed a decrease in passengers for the fifth year in succession, although goods traffic had continued to grow. The finances of the company had been impacted again by the need to undertake repairs to one of the locomotives, £200 spent on an overhaul of the engine.

The company, despite the tumbling passenger figures over the past five years, decided early in 1912 to invest in two new passenger cars at a total cost of £300 to replace the unsuccessful Reading cars. The cars were both built by Hurst, Nelson of Glasgow and were converted double deck tram cars. The upper deck was removed, and a new roof placed on the car to complete the conversion. The cars were the first to run on bogies on the tramway and were luxuriously upholstered, which no doubt improved the quality of the journey immensely. The two new cars were given the Nos. 4 and 5 which had been carried by the Reading cars before their withdrawal and scrapping.

Some good news was received early in 1912 when the post office announced that it wished to extend the contract with the tramway company for the carriage of mails. All the post office traffic in 'letters and parcels, bags and boxes' was to continue to be

Passenger car No. 4 undergoing conversion at Hurst, Nelson and Company's workshops in Motherwell. Originally built as a double deck tram, the car was converted to a single deck car for the Wantage Tramway. The stairs to the original upper deck, which has been removed, can be seen in the doorway. *John Scott-Morgan Collection*

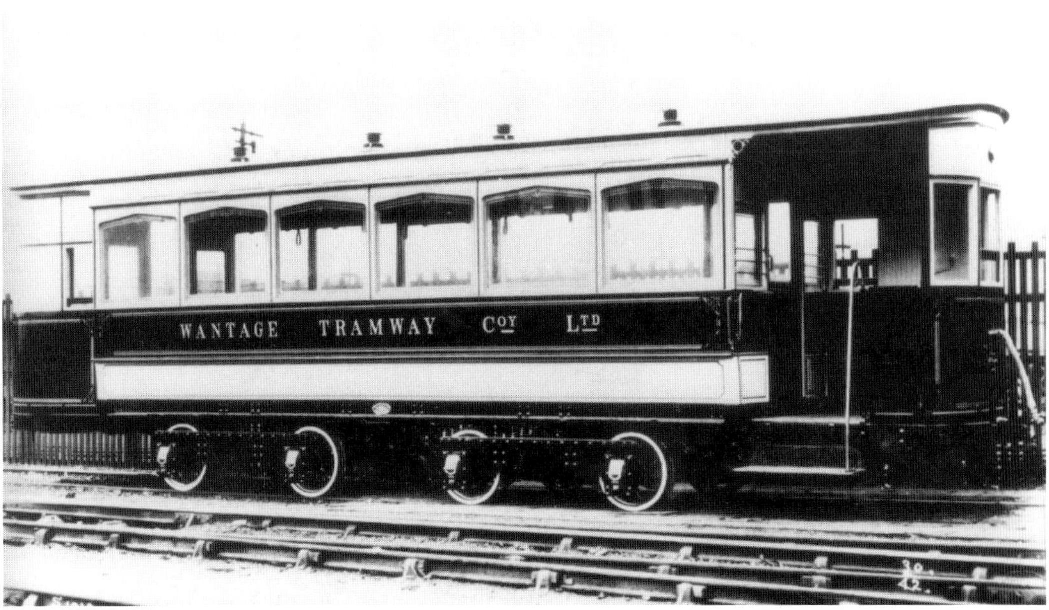

Passenger tram car No. 4 when new, having just been delivered to Wantage Road yard by the GWR. *Vale and Downland Museum, Wantage*

The interior of the abandoned passenger car number 4 showing the luxuriously upholstered longitudinal bench seats. The door at the far end leads out onto the end balcony of the car from where passengers enter or alight from the car. The white poster above the left window is a notice prohibiting smoking in the tramcar. *John Scott-Morgan Collection*

transported from Wantage Road by the tramway at an annually agreed rate of £75 6s per year. The contract was to be extended for five years.

1912 seemed to be a generally poor year for the tramway. Passenger traffic continued to decline, and both trade and heavy goods tonnage fell. The decline across all of the tramway's traffic was blamed upon the national coal miners' strike, the first time the coal miners' union The Miners' Federation of Great Britain had called for national industrial action. The strike had lasted for thirty-seven days between 28 February and 6 April with no agreement being reached between the miners and the private colliery owners. With no consensus among colliery owners, and long before the advent of the National Coal Board, the strike seemed to be destined to continue, until the government intervening by passing the Coal Mines Act, which extended minimum wage provisions to not only the mining industry but all industries which had many manual jobs, ended the dispute. The strike caused a large amount of disruption to train and shipping services due the lack of steam coal upon which these industries relied. The tramway would have been hit particularly hard as one of its main sources of heavy freight traffic was the coal coming in to Wantage from Wantage Road for the gas works and town's industries.

Despite the difficult trading conditions the company had returned a healthy profit, enabling the directors to declare a shareholders' dividend.

With the benefit of hindsight, it becomes clear that the tramway's problem which led ultimately to the decision in the 1920s to terminate the passenger service had begun long before the First World War. Passenger numbers had been in terminal decline since 1907 and remained so in 1913, although only declining slightly over the year when compared with the passenger receipts in 1912. Goods traffic had been relatively stagnant during the year, with the final tonnage statistics showing a very slight decrease, but only by a few tons.

The tramway company had, however, remained profitable, returning an annual surplus of £465 2s 10d, part of which was assigned to the reserve fund, the remainder being used to pay the shareholders another dividend on their shares.

The directors and shareholders of the company had no idea what was to occur in 1914, although many of them would have been aware of the worsening political situation in central Europe as the Austro-Hungarian and Ottoman Empires, Greece, Bulgaria, Romania, Montenegro and Serbia fought a series of Balkan Wars during 1912 and 1913. The worsening political relations after the Second Balkan War left Austria-Hungary and Serbia vying for position in Central Europe, which would spiral into open conflict in the new year, but no one in Wantage would have been able to foresee the global war that was to come.

Chapter 10

THE WANTAGE TRAMWAY DURING THE FIRST WORLD WAR

The new year of 1914 dawned with the storm clouds that would gather still far off on the political horizon.

The end of the financial year on 30 September 1914 saw the First World War having little affect upon the tramway. Hostilities had only broken out between the United Kingdom and Germany on 4 August, with declarations of war against the other two central powers Austria-Hungary and the Ottoman Empire being given on 12 August and 5 November respectively. The directors' report stated that the tramway's year had been to the directors' expectations both operationally and financially until the end of July. The report states that all goods traffic was practically at a standstill during August. This may have been due to the mobilisation of Britain's armed forces following the declaration of war and the despatch of the British Expeditionary Force to France, as well as the Royal Navy preparing for an anticipated large naval battle with their German counterparts.

The company remained profitable despite the global upheaval, returning a heathy profit of £445 8s 4d, most of which was again used to provide good dividends for the shareholders.

The tramway did not have a particularly good year in 1915. The directors' report to the shareholders showed that passenger receipts had been slightly down on the returns for 1914, but goods traffic was very much less than previously, coal in to the gas works and town with flour outbound from the mill in Wantage and iron and steel products for the armed forces from the iron works the only real traffic that had been available to the tramway. The company reported that all goods and parcel traffic had been 'greatly disorganised' by the needs of the military during the financial year, and that by the end of the September 1915 very little non-military traffic had been moving on the railways. Construction materials had not been carried at all, the traffic completely stopping due to the building trade being stopped by the war, the construction of naval facilities, army buildings and static defences probably taking up most of the materials and labour available to the construction industry.

The tramway's loss of goods traffic was contributed to by another industry which had not been operating at all. Horse racing had been completely stopped during 1915. The Berkshire downs was usually a centre of horse racing, with many stables located around Wantage and Lambourn. In 1915, all of the horse racing establishments had been closed, the closure of the racing stables at Letcombe Bassett and Letcombe Regis particularly affecting the tramway's goods traffic, since no race horses were travelling to race meetings, most of which began their journeys at Wantage, the horse box vans being exchanged with the GWR at Wantage Road both outbound and upon return. The racing season simply had not happened with a consequent loss of the revenue to the Wantage Tramway.

The war had also caused the prices of coal, corn and raw materials to increase. Coal was being purchased in huge quantities by the Royal Navy since the majority of its warships were still coal powered at the time. Corn imports had become more expensive due to the increased cost of shipping due to military use of freighters, and the U-Boat threat around Britain. Raw material costs had also increased due to demand for them being increased by war production. As commodities, all of these items were increasing in price as demand drove up prices.

The tramway as a result of the increasing cost of fuel and living, increased its carriage charges for the next financial year to absorb the increased energy costs. The company has also reduced its staffing levels to the bare minimum, laying off some of its employees. Those who remained were paid a war bonus though.

Despite its commercial difficulties for the year, the tramway still made a profit, although the £23 7s 10d was insufficient to allow the directors to declare a shareholders' dividend. The 1915 financial year was, as a result, one of only a few years when the tramway shareholders did not receive a return on their investment. The new financial year offered a potentially better return, as freight traffic had already begun to recover during late August and into September 1915.

1916 saw the company return to normal trading as traffic began to recover from the shock of the opening of hostilities across the globe when the war started. The cost of coal and wages had continued to increase during the financial year. Passenger traffic had, unsurprisingly, continued to decline, although this was indicative of the trend which had begun in 1907 rather than the impact of the war. Freight traffic had increased dramatically compared to 1915, with both trade and heavy goods returning to prewar levels. Parcels traffic had increased over the year exceeding pre-war levels.

The company, now busy once again, began to look to rehire some of the workforce it had laid off during 1915. Profits for the year amounted to £257 15s 6d, despite increased costs, which allowed to the directors to add to the reserve fund and also to declare a dividend for the year.

The 1917 financial year was less profitable than the previous year. The flour mill in Wantage had temporarily closed with a loss of between £700 and £800 in traffic revenue for the tramway company. The loss of the flour traffic almost halved the heavy goods returns for the year, with carriage charges contributing £510 4s 2d. Parcels traffic for the year was also very low, only £31 19s 5d added to the company's funds, although the traffic was an increase on the previous year's parcels income. The annual report reported a little good news for the shareholders, passenger traffic having increased by a small amount with fares amounting to £97 15s 3d.

The company had reduced its costs during 1915, and even with the rehire of a couple of junior employees during 1917, wages costs were still low at £57 19s 6d. The directors of the company took no salary from the company in 1917, deciding to abstain from their entitled payments due to the difficult trading conditions that the company was experiencing.

The company announced that it was once more increasing passenger fares and haulage rates for goods and parcels due to the increasing costs of fuel and wages. The announcement softened the news that the shareholders were not to receive a dividend for the 1917 financial year since the company's profits had been minimal at only £162 2s 4d.

As the year drew to a close, Edward Ormond, who had been involved with the Wantage Tramway since the very beginning as both promoter and company secretary, resigned. He wished to retire both as a director of the company and as company secretary due to his advancing age. His resignation was regretfully accepted. Ormond's son, Edward Brooks, had been suggested as a possible successor, and since he held the required twenty shares in his own right, he was eligible for a directorship. The directors accepted the recommendation and duly elected Edward Brooks Ormond to the board and, with his also being a solicitor, asked him to take over as company secretary which he accepted.

In 1918, the tramway was running an improved service of nine trains each way every day except for Sunday when no service was run. The trams were timed to meet GWR trains at Wantage Road, with the first tram of the day leaving Wantage Town at 07.00 to meet the 07.26 Up and 08.05 Down trains, with passengers arriving on the earlier Up train having to wait for any passengers off the later train to join the tramway train. Trains were timed irregularly during the day, always timed to meet a GWR stopping train. The other eight trams each day left Wantage Town at 09.20 to meet the 09.43 Down train at Wantage Road, at 09.40 to meet the 10.07 Up, 13.05 for the 13.29 Down, 14.25 for the 14.50 Up, 17.20 for the 17.47 Down, 18.15 for the 18.42 Up, 19.35 for the 20.01 Up and the last of the day leaving Wantage Town at 20.25 to meet the 20.52 Down GWR train.

The tramway company also introduced a special service which would run at 07.40 just to meet the 08.05 train for a minimum of four passengers upon a request being made at the tramway offices by the previous evening. All trains, including the special service, aimed to leave Wantage Road as soon as possible after the arrival of the main line train service but not to an advertised timetable. A special allowance had been introduced during 1918 for commercial travellers on all tramway passenger services. All passengers were able to travel with luggage up to a maximum weight of 56lb without charge, but commercial travellers were to be allowed an additional 56lbs of luggage for free.

Fares were publicly displayed at Wantage for all trips on the line, with a full line journey costing 9d, travel between Wantage and Grove Bridge (3d), Wantage and Oxford Lane (6d), Wantage Road and Oxford Lane (3d) and Wantage Road and Grove Bridge (6d) all being available to purchase from the conductor onboard the tramcars. Children under 3 travelled for free, with those under 12 travelling at half fare. Special fares were available from Wantage Road on market day on Wednesdays, with a full line ticket costing 6d, Oxford Lane to Wantage costing 4d and Grove Bridge to Wantage costing 2d. No return tickets were available, a ticket having to be purchased for each journey taken, although a special ticket had been introduced prior to the First World War for anyone accompanying another traveller to Wantage Road station to see them off on a GWR service. The return 'seeing off' ticket cost 9d.

Goods services were carrying an increasing amount of freight, with the three goods trains each way consisting of between six and twenty trucks, with an average train being made up of twelve wagons. The trains travelled at no more than 6mph and so no guards van was ever used, which would result in a fatal accident.

Manning Wardle No. 7 with passenger cars 5 and 4 at Wantage Road station. Note the Great Western Railway's yard crane in the background. *Vale and Downland Museum, Wantage*

The tramway was involved in a fatal accident in 1918 when a 6-year-old boy was killed on the line. Cyril Adams was run over and fatally injured by the tram on 24 September when he ran out of a field on the west of the line opposite Canal Cottages, accompanied by two other boys, at around five o clock in the evening. The approaching train was a goods service made up of seventeen wagons which was heading towards Wantage from Wantage Road proceeding at just 4mph. Ropes had been used to secure loads on the wagons as was normal procedure and the ends were hanging from some of the wagons. Adams and one other boy 'caught hold' of the ropes on the fourteenth wagon and whilst hanging off one of them, Cyril slipped and stumbled and fell onto the rail and was run over by the following wagons, dying of his injuries shortly afterwards.

The incident was reported to the BoT who assigned Colonel Pringle to investigate. It was ascertained that the driver had not seen the boys and despite keeping a good look out had no way of seeing what they were doing. The driver was found not to have been responsible for the accident in any way. It was however suggested that in the case of long goods trains such as the one involved in the accident 'a competent person should ride at the end of the train to act as brakesman and lookout'. The byelaws of the Wantage Tramway, made by the BoT when the tramway opened, did not require the use of a brake van on goods trains as only the operation of passenger trains had been covered in the byelaws. Pringle's report stated that it was unusual for trains of such a length to be running on tramways without a guard's van attached to the rear, no other cases of such trains having been reported on any other tram line at that time.

Pringle carried out a general inspection of the tramway as a result of the incident and whilst the tramway was found not to have been guilty of any breach of rules or of incompetence that led to the accident, recommendations were made to avoid a similar incident. The drivers were to be reminded of the importance of keeping a good lookout at all times, and of the necessity to obey the byelaw requiring all engines to be stopped before the level crossing at Grove Street which was occasionally being ignored. Five of the intermediate crossing places had been seen to be no longer in use, with Pringle reporting that the padlocking of the points levers was insufficient, the points were also to be clipped to prevent them opening under a train. The general state of parts of the line was commented on in the report, which although generally good, required some remedial action at places where trees and hedges adjacent to the line were growing over it, and also at Wantage Town yard where ash and general refuse left on the ground was viewed as a potential hazard to staff during shunting.

The last recommendation concerned the manner of operating the line which had been observed during the inspection. The trams were leaving Wantage and Wantage Road with no formal system of control in place to safely despatch them. In future, under a new agreed byelaw trains or engines travelling light were only to be despatched from the terminuses if no other train had been sent from the opposite

No. 5 shunting at Wantage in the upper yard. The chain attached to the locomotive's coupling hook is being used to fly shunt wagons in the constrained yard. The lower yard had much more space and fly shunting was not needed. *John Scott-Morgan Collection*

terminus. If two trains had already been sent, then a train could only depart once both had arrived from the other end of the line. Verbal arrangements were to be made where a train must receive permission to proceed from a duty manager at the terminus. A telephone system was to be installed between Wantage Town and Wantage Road station to allow for the control of trains.

The BoT inspectors report stated that the Wantage Tramway was effectively operating as a light railway rather than a simple tramway and so recommendations were made to the board that the byelaws in force for the tramway company needed to be updated to reflect how the operation had evolved since it had opened. The BoT issued an additional set of byelaws which the tramway implemented immediately. The new regulations limited goods trains on the tramway to be made up of no more than fifteen wagons, with a brakevan required on each goods train. Goods services were only to be allowed to run at 4mph, and a guard or conductor was required to be present in the guard's van at all times when the train was moving.

The financial year ended on 30 September before the war ended, but whilst the final Allied offensive of the war, later referred to as the '100 Days Offensive' was progressing well against the German army. The war elsewhere was drawing to a close, with British and Italian forces pushing the Austro-Hungarian army back in northern Italy, whilst British and French forces pursued the withdrawing Ottoman army in the Middle East. The directors' report for the year echoed the optimism that was being felt across the whole of the country, but presented a mixed picture with respect to the tramway's operations during the year.

The company's income from freight traffic had declined during the year, only £153 7s 0d of carriage charges for goods being paid during the year. Parcels traffic fared little better, declining to just £25 16s by the end of the financial year. Passenger traffic had increased dramatically, however. Fares for journeys along the tramway had leapt to £1,619 5s 4d, although wages had increased to £188 for the year due to both an increase in staffing levels and rising pay, which was determined by the government's wages council. The company ended the year with a healthy profit of £681 11s 1d, which allowed the directors to announce a good dividend to the shareholders.

1919 began with the company's locomotives worn out after hard work during the war years and in early March it was decided to advertise for a light steam engine for hire. The company approached Swindon to establish if they had room in the work's schedule to overhaul one of the company's locomotives in the near future, whilst also writing to Manning Wardle to ask them if they would be able to overhaul another of the tramway's engines. Manning Wardle were able to accommodate one of the tramway's locomotives, and in early April No. 5 was sent to Leeds for a full overhaul.

The company changed its approach regarding an additional engine in the spring, searching for a second hand locomotive to purchase. A.J. Belcher accompanied the tramway's manager to Woolwich Arsenal in early June 1919 where they inspected all of the locomotives that were to be sold. One of the engines was deemed to be suitable and the tramway bid £600 for it, winning the tender and becoming the new owners of the engine. Arriving at Wantage the company's new machine named *The Driver* was put into service, but only briefly.

The tramway's drivers reported that the engine's brakes were worn out and needed to be completely replaced. The locomotive's boiler was then found to be in a terrible state of repair and it was parked up at Wantage whilst the company's engineer, Wilkins, who had recently been promoted to the post, undertook a complete inspection of the engine.

The bad news continued with driver Weaving informing the company's manager that he had found better paid employment as a driver elsewhere and would be leaving although if the company were prepared to match what was on offer from the other company, he would prefer to stay. Weaving's case was referred to the tramway's directors who authorised an increase to his pay from 35s to 40s per week, with overtime pay to be increased from 6d to 1s per hour extra. Weaving accepted and remained with the tramway.

As the winter approached, one of the company's other drivers announced that he had decided to retire. John Hewitt, who been driving the engines for the tramway for 45 years, left the company at the end of the company's trading year on 30 September 1919 with the best wishes of all of his colleagues and the directors of the company. He would not be away for long.

The 1918-1919 financial year had once again seen good passenger numbers and increasing amounts of goods traffic. Passenger fares had risen to £1,863 5s 1d, probably as a result of more travellers returning from military service during the recent war. Goods traffic returned £2,539 16s 1d, with parcels adding another £205 14s 1d to the company's finances. Wages for the year had been very high though, with £2,075 13s 4d spent on pay for the company's staff. The company remained very profitable for the 1918-19 financial year returning a surplus of £730 6s 9d, which allowed the company to pay another good dividend to its shareholders. The results of the previous two years and the huge increase in passenger receipts gave the directors and shareholders cause for optimism that the passenger service would, as they expected, return to being well patronised. The future would be quite different.

Despite a very slight recovery during the war, and the huge increase in passengers in the 1917-18 and 1918-19 financial year, passenger traffic would not be sustained as the tramway went into the 1920s. The huge increase in passengers was most likely to have been due to service personnel travelling on railway warrants which meant they had to use the tramway when they got to Wantage Road. The same was likely to be the case during 1919 and early 1920 as service personnel were demobbed and returned home, travelling once more on military railway warrants. The 1920s would see the passenger service struggle for passengers.

One of the first tasks for the new financial year was to resolve the defects with the engine that had been purchased from Woolwich Arsenal. The manager William Noble was instructed to write to them to complain about the state of the engine that had been sold to them. It had been found to be in a bad state of mechanical repair, which could not have been observed when it was inspected in Woolwich and the boiler tubes were also defective. A letter was sent to the arsenal at the beginning of October, but no reply was received from Woolwich. The manager wrote again on 3 November and still received no reply.

Passenger traffic had seemed to be very good in 1918 and 1919, and so in November the company directors decided to increase passenger fares from 9d to 1s as from 1 December. The directors had not realised that the passenger traffic increases were only temporary due to the end of the war. The increase in cost made the service extremely expensive and would accelerate the decline of the tramway's passenger traffic in the early part of the next decade.

A major issue began to present itself to the company from late November. At the directors' meeting on 3 December the case of the missing manager was discussed. Noble had not been to work since 24 November, and nothing had been heard from him, and no one in Wantage had seen him. In the absence of a manager, one of the directors J.A. Webber offered to supervise the operation of the tramway until a new manager could be appointed. His offer was accepted, and it was agreed to pay Webber a salary of £200 per year to undertake the task. The board resolved to dismiss Noble as from the last date he had attended work, but little did they know of the storm that was about to hit the company. Upon inspecting the company's books and with the other directors realising what had been transpiring, Webber withdrew his offer at the 12 December directors' meeting. It was decided that in view of the problems that had been found and the state the tramway had been left in following the disappearance of Noble, an extraordinary general meeting of the company's shareholders should be called to 'consider the company's affairs', which was scheduled for 1 January 1920.

Chapter 11

INTO THE 1920s

The new decade began badly for the tramway. It was found that the company's manager, William Noble had absconded late in 1919, with a large amount of the company's money, which had been in his care. Noble had also been keeping money out of the company's takings when the fares boxes had been dropped off in his office, as well as not stating in the books all of the freight traffic which the tramway had carried. He then kept the payments when they came in for himself. At the end of 1919, Noble had disappeared and taken his ill-gotten gains with him. The directors soon found out what had happened and began an investigation which revealed the truth. The goods manager's traffic ledgers were found to differ from the general manager's payment ledgers, the extent of Noble's theft becoming clear, and also seriously threatening the company's future with cash not in the bank to pay wages and bills. Effectively the tramway had been running at a loss whilst Noble was filling his own pockets.

At the extraordinary general meeting of the directors and shareholders that had been called in response to the situation, the directors reported to the gathered shareholders that they had called in the company's auditors to go through the company books forensically to establish the extent of Noble's theft from the company. The auditors' interim report had revealed that there were serious shortcomings in the way in which the company books were being kept, and how cash was handled. The working practices in the company's offices, which had been suitable in the past but had not evolved as the company grew, had given Noble the opportunity to steal from the company. The auditors' early examinations of the company accounts had found 'grave discrepancies' in the books suggesting that Noble had stolen a large amount of money from the company. The meeting agreed with the directors that the matter should be referred immediately to the police, and the company should seek to prosecute the erstwhile manager for embezzlement. A second extraordinary general meeting was scheduled for 29 January, by which time the auditors should have completed their investigations. The final act of the meeting was to authorise the company secretary Edward Ormond to agree to the issue of a warrant for the arrest of William Arthur Noble as recommended by the police officers investigating the case.

The second meeting went ahead on 29 January as agreed, the auditor's report making uncomfortable reading for all present at the meeting in Wantage. The auditors had found that the accounts had been irregularly kept, and that the present system of managing the company's finances in the offices was unsuitable. The current accounting system had provided an opportunity for Noble's embezzlement. It had also been found that a large proportion of the outstanding amounts due to the company were debts which should have been dealt with by the directors a long time ago. It seemed likely that Noble had kept some of the payments and not marked them as paid in the ledgers as well as understating daily receipts for passenger fares, parcels charges and upfront payments for goods traffic, if the customer had paid the amounts to him rather than the goods manager.

The auditors' report made a number of recommendations. The current directors were to be asked to retire, and new board of five directors should be elected to replace them. Secondly, an auditor experienced in railway accounts ought to be appointed as soon as possible. The tramway company was advised to appoint a new manager on 'such a salary as will secure a good man', it being suggested that the salary being paid to Noble had not discouraged him from stealing from the company. The company was to adopt a new accounting process which would make it impossible for a similar incident to occur, as well as revising the system of customer credit. A full stocktake of all of the company's assets and property was to be taken annually from the end of the 1920 financial year.

The auditors suggested, which the shareholders, managing engineer and directors agreed with, that the company should ask the GWR for assistance in inspecting the rolling stock and line as it was believed that the erstwhile manager had been skimping on maintenance which the engineer and directors believed had been completed to a satisfactory level and to their instructions.

The annual general meeting was scheduled at the second extraordinary general meeting of the company, for 10 February. The directors of the company agreed to retire at that meeting, with five men expressing their wish to be elected to the new board at that meeting. The new board was duly elected, with five new directors appointed. Dr W.D. Loveday was elected as the new company chairman, with A.E. Denis de Vitre, P.J. Clark. E.J.W. Walden joining him on the board. It was agreed that Edward Brooks Ormond should remain on the board and continue as company secretary since he had only been elected to the previous board at the end of 1917.

Ormond had written to Wake and Company in Darlington, during the previous December, to ask if they would be interested in purchasing two locomotives, then out of use in the upper yard at Wantage for scrap. Having received no reply, Ormond wrote them once more in early February 1920 asking Wake's if they intended to tender for the locomotives, to which a reply was received that they would. Ormond also wrote to the GWR, addressing a letter to the company's general manager which 'called his attention' to the inadequate compensation being paid to the tramway for the collection of money due to the GWR for deliveries of goods which received no reply.

The tramway hoped to have a new manager in place relatively quickly, with many applicants responding to the company's advertisements. Four applicants were invited to come to Wantage for interview, with the company offering to pay their third class rail fares, and expenses. Three of the men accepted and attended the interviews which were arranged for 2 March. H.A. Turner, then a stationmaster on the GWR, was offered the position at the generous salary of £250 per year, which he accepted. It was agreed that Turner would take up the position in one month.

A telegram was received the following day from Turner stating that he had changed his mind and no longer wished to accept the appointment. Henry Edward De Vitre, one of the tramway's directors travelled to see Turner at his home, and Turner accepted the post on an increased salary of £275 per year for the initial six months, with a raise to £300 per year after that. The company finally had a new long term manager, or so they thought.

The company secretary finally received an offer from Wake's for the two redundant locomotives which was deemed too low and rejected. Another offer for all four of the engines out of use at Wantage was received from John Cashmore and Son of Great Bridge Staffordshire which was deemed to be acceptable. Cashmore offered £520 for the 'three old engines' and £120 for the fourth with an additional £6 per ton for other scrap metal to be disposed of. The three 'old' locomotives out of traffic that the tramway wanted to dispose of were the Hughes Tram, No. 4, the locomotive which had been purchased from Woolwich Arsenal and proved to be such a disaster, and the old GWR tank engine 1329 *Raven* which had been out of service since being badly damaged in a collision on the tramway. The fourth engine, for which Cashmore had offered £120 was No. 5.

A visit from an engineer from Swindon and representatives from the company's boiler insurers occurred in late March 1920, during which the locomotives were all assessed to establish which could economically and quickly be repaired. The Woolwich engine *The Driver* was viewed as needing a complete rebuild which would take six months and cost a considerable amount of money. With only the Matthews tram engine and No. 7 in service, the company needed to quickly find a third engine to cover both the passenger and goods traffic and could not wait for the Woolwich locomotive to be repaired at Swindon. No. 5 was inspected and found to be in better condition and could be quickly repaired by the GWR, and the company accepted the engineer's report and decided to retain the little George England tank engine. The other three engines were dismantled for scrap in the middle of April.

With the motive power situation resolved, the tramway approached the GWR offering to undertake all of the shunting in the yard at Wantage Road for a weekly payment of £10, which would save the company from having to send an engine from Didcot to do it. It seems the approach was not replied to.

A rear view of No. 7 with a goods train near the lower yard in 1944. The square panel underneath the buffer beam is the ashpan damper used to let air in to feed the fire through the firebox grate. *John Scott-Morgan Collection*

As the summer began, the company received an approach from a party who expressed an interest in possibly purchasing the tramway. A series of letters were exchanged and with discussions proceeding well it was agreed to send the company accounts for the previous four years to the gentleman concerned. The accounts were duly prepared and packaged up for delivery to Colonel Holman F. Stephens. Stephens was a civil engineer and owned many light railways all over Britain and has more recently become something of a railway legend. The accounts were seemingly acceptable to the Colonel as he requested that his inspector be allowed to walk the line and inspect it. Permission was granted on 29 June with the inspector arriving shortly afterwards.

It seems that all was not well on the line in 1920. A major setback was experienced when on 7 July the tramway's manager H.A. Turner resigned and left the company. In order to provide direction for the ailing tramway, H. Denis de Vitre was appointed managing director, the company's first, on a salary of £250 per year. The new managing director promptly set about the process of turning the fortunes of the company around. A group of employees received pay rises and other improvements were made all over the line. A new cadre of office staff were recruited, with James Welch joining the company as 'checker of goods and parcels' on a wage of £2 per

The Matthews Steam Tram engine No. 6, with car 3 at Wantage some time in the 1920s. The men gathered around the train are all employees of the company but unfortunately no note of who they are was kept. *John Scott-Morgan Collection*

week. He was joined by Stokes, the new managing clerk (£4 per week) and Miss Payne as goods office clerk (£1 8s per week) all of whom were in place by 28 July.

In early August 1920, the tramway received a letter from Colonel Stephens expressing an interest in purchasing the line. The directors agreed to meet the Colonel on 20 August. It seems though that an agreement could not be reached as no further communications between the Colonel and tramway exist in the correspondence files, and the tramway remained in the hands of its shareholders. Colonel Stephens appears to have ended his interest and the Wantage Tramway did not become part of the Colonel's light railway empire.

In late August, the tramway was once more short of motive power, with No. 5 in the shed at Wantage undergoing repairs by Wilkins. The company attempted to purchase an engine to bolster the fleet but without success. An approach to Ward's foundry in Wales seemed to result in the company finding an ideal locomotive but having learned from their experience with the Woolwich engine delayed purchasing it until an inspector from the company's boiler insurance company had reported on it. The report was not complimentary, and the company decided not to proceed with the purchase. Wake's in Darlington offered the company a locomotive that they had purchased for scrap but that was in reasonable condition. The engine was expensive, Wake's wanting £2,250 for it. The company asked for a better price but also for them to delay selling or cutting up the engine as the tramway would be able to purchase it in the near future.

The company set about raising money to buy Wake's engine, and by the end of November had a commitment from A.K. Lloyd for a loan of £1,500 in return for 7 per cent interest debentures, whilst Ormond had agreed to lend the company a further £200 on the same terms. Whilst the funds were being raised, the company had been reviewing the state of the engine which was sitting dismantled in their shed at Wantage. It was decided to send the engine to Swindon for repair and to delay purchasing a fourth engine, and with a view to arranging the overhaul, Dr William Loveday, and Phillip Clark, who had taken over both his father's milling business and directorship of the tramway after the latter's death, travelled to Swindon to meet the GWR's Chief Mechanical Engineer George Churchward. Churchward was willing to help the tramway, and agreed to provide an estimate and a final decision as soon as he could. He wrote to Dr Loveday on 16 December agreeing to undertake the repairs to the tramway's locomotive as a priority and the engine was sent to Swindon.

The company continued to struggle with its cash flow following the theft of its funds by William Noble. Whilst the company managed to survive the 1919-20 financial year through good management, the financial position of the tramway was becoming dangerous by the end of 1920. It was decided at the 16 December directors' meeting to ask the shareholders at the next annual general meeting, scheduled for early in the new year, for permission to raise funds of £3,000, either through a debenture issue or a mortgage against the company's property. The tramway was also informed that Edward Brooks Ormond's father, one of the original proposers whose work had been one of the major contributors to the tramway being built, had passed away earlier in the month. Edward Ormond had left his all of his ordinary and preference shares to his son in his will. Edward Brooks Ormond was now one of the major shareholders in the company.

1921 began with the tramway recovering well from the trouble caused by their erstwhile manager, Noble's mismanagement of the tramway's permanent way remaining the last outstanding problem to resolve. With funds now flowing into the company's accounts from the line's traffic, and profits from the previous two years allowing the management to invest, work on repairing the line was once more possible. The tramway ordered 12 tons of second hand rails and fittings from the Midland and South Western Junction Railway at the cost of £15 per ton. As soon as the materials arrived, the tramway's engineer and maintenance staff set about making good the line.

A long ongoing discussion between the tramway directors and Philip Clark, also a director but in his capacity as owner of Clark's Flour Mill, ended in the early part of the year. The negotiations had involved the charges being made by the tramway for the traffic that the mill was placing with the company. The company had offered a rate of 3s 3d a ton for all freight, with Clark making a counter offer of 2s 6d. The tramway then offered a sliding scale of charges, which Clark refused. On 22 January, the directors of the tramway, reluctant to lose the traffic agreed to settle for Clark's offer.

The summer provided the tramway with an extra source of income. The trees in the orchards owned by the tramway and no longer occupied by any tenants produced a huge crop of apples. The orchard's crops had previously been sold to raise revenue for the tramway whilst no tenant was in place, and in 1921 the bumper crop was to be sold by auction, raising £26. The company was about to lose a major regular source of revenue though.

On 27 July, Philip Clark informed the tramway that the mill was serving notice of their intention to terminate their haulage agreement with the line, and so the necessary six month notice was thereby served. The same day, the chairman, Dr Loveday informed the company that he would be leaving Wantage and was obliged to resign as both a director, and chairman of the company. Loveday continued as chairman until a suitable replacement expressed an interest in being elected, and in October Ernest Nalder offered his services and was duly elected. Dr Loveday stood down at the same meeting, ending his directorship of the company on 5 October.

A work stained No. 5 outside the original running shed at Wantage. The canvas curtain seems to have been left closed, suggesting that the weather may have been poor. *John Scott-Morgan Collection*

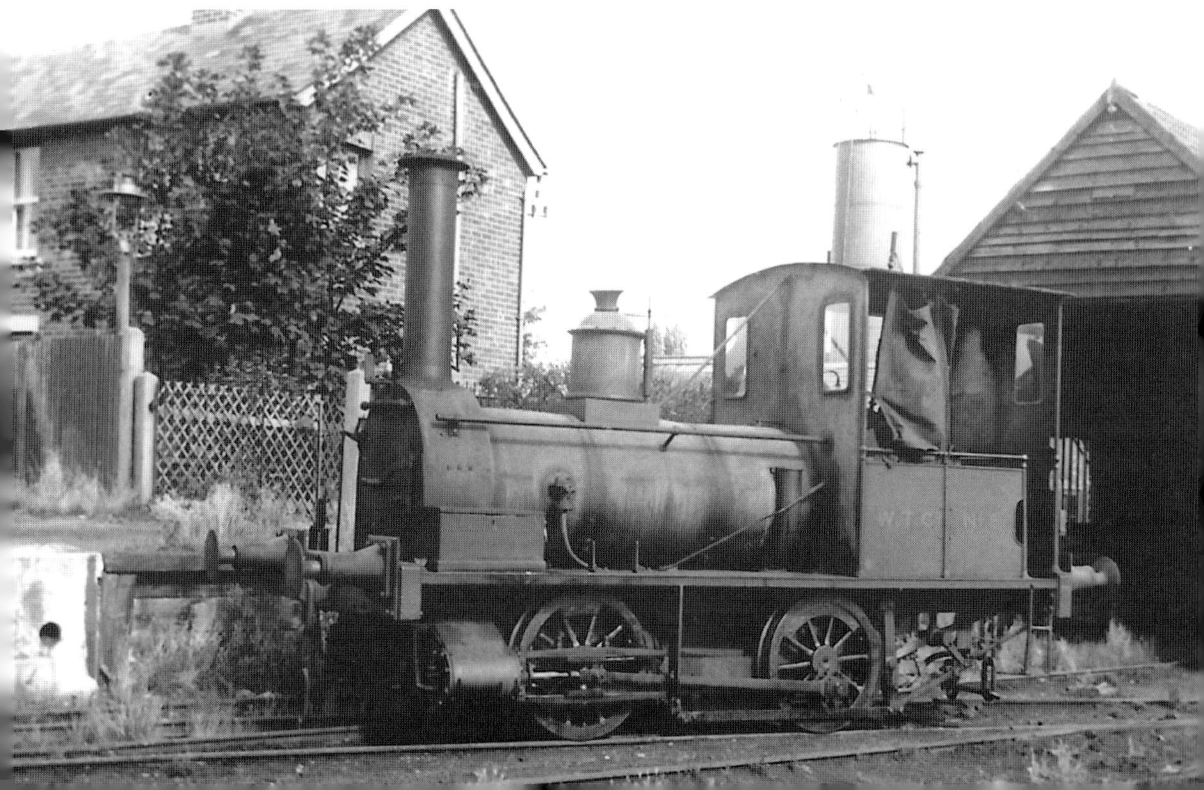

The end of the month brought mixed fortunes to the tramway company. No. 5 had returned from Swindon following repairs and was back in service, resolving the motive power shortages. The three locomotives took turns on passenger and freight services, although the Matthews tram only worked passenger services. The company seemed to be recovering well and was optimistic about its fortunes. The bubble burst when the tramway was confronted with theft by employees again.

The company's managing clerk, Stokes, had absconded with £14 4s 11d of the company's cash. Upon investigation it was found that Stokes had actually stolen a considerable amount more, £201 4s 3½d found to be missing. Legal action was taken against Stokes to retrieve the missing money and the police were informed.

At the November directors' meeting, Denis de Vitre offered his resignation as managing director, but offered to stay on until a replacement was appointed. Applications for the post of manager were reviewed and a new manager was appointed on an annual salary of £325, to start in the new year. The search for a new manager ended when Joseph Bullock, formerly the company's goods manager, was appointed.

Bullock's first task was to resolve a concern which has arisen with one of the company's drivers. Many complaints had been received about the unsafe driving of the passenger services and general lack of attention to his duties whilst driving on the part of C.A. Clarke. The new manager was left with little option other than to dismiss Clarke from the tramway's employ. Clarke was not prepared to accept he had done anything wrong, and wrote to Bullock on 1 February. He asked for an explanation of the reasons for his dismissal, but the company replied that they were not prepared to enter into discussion or to give reasons for Clarke's dismissal, other than it was in the best interests of the company for his employment to be ended.

The Matthew's tram was out of service during the late spring, requiring a new firebox. The company had also had to send Nos. 5 and 7 away to Swindon for repairs, which when added to the cost of repairing No. 6 presented the company with a large and unexpected bill to pay. The company directors approached the GWR who agreed to allow the tramway to pay the bill in instalments. The Great Western once more seemed keen to support their little neighbour. A cheque was sent for £471 as payment of the first part of the debt, with another £500 to follow on 1 September and the final payment to be made on 1 January 1923. The full amount to paid included 5 per cent interest applied to the debt by the GWR.

The financial constraints that had been placed on the company resulted in the need to reduce costs temporarily, and as part of the necessary economy drive, all wages and salaries across the whole company, including his own salary, were reduced by Bullock. The economies made, and an agreement from the company's bank to allow a short-term overdraft allowed the tramway to begin to recover from the unexpected repair costs. The position was helped when the GWR informed the

No 5 often hauled the tramway's passenger services, and is seen with passenger cars 4 and 5 on a Wantage bound tram. When the bogie car No. 4 arrived on the tramway, it was noticed that the entrance to the car was at the end of the platform away from the waiting room at Wantage, but that the other entrance was at the opposite side of the car from the platform. It was sent to Swindon to be turned and when it returned the diagonally opposite entrances had just switched places and the problem remained ! *John Scott-Morgan Collection*

tramway that it was prepared to waive its interest charges if the two payments were made on time at the dates agreed.

Early October saw the company receive an application from the Anglo-American Oil Company, who leased a parcel of the tramway's land at Wantage where it had built storage tanks for oil. The oil company's request was granted, and a new lease was drawn up for a period of five years at an annual rent of £10. The tramway and oil company also agreed that the carriage rates for the oil which was carried on the line would not be increased, and would remain at 2s 6d per ton.

The tramway continued to have accidents and near misses involving road users on the Besselsleigh Road during the 1920s. No. 5 in particular appeared to be

Wantage Tramway No. 5 in the yard at Wantage Town. Note the yard crane behind the locomotive which was purchased from the Wilts and Berks Canal Company. The tank wagon behind the engine is one the Anglo-American Oil Company's vehicles. *John Scott-Morgan Collection*

unfortunate, being involved in a few accidents and incidents during 1922. A letter to the tramway, dated 3 December 1922, reported that the locomotive had collided with a motor car and a cart in two separate incidents and had run over and killed a bullock on the road. The permanent way was reported to appear to need maintenance in 1922, with the rails at least needing to be cleaned out and levelled off. No. 5 was reported to have derailed six times during the year. The GWR staff at Wantage Road had begun to despair when the telephone rang to report another tram off the rails on the tramway, a loud call of 'the tram's off again' becoming the station staff's usual means of alerting the station master that alternative transport to Wantage would need to be laid on for passengers arriving on GWR services stopping at the station.

1922 saw the tramway company in a legal dispute with the High Court of Justice on 6 February. The company had been summoned to answer a charge of having not correctly registered a mortgage loan which had been taken out between the tramway company and John Arbery for the sum of £600 on 12 June 1919. The company had failed to register the mortgage within the time allowed by the Companies (Consolidation) Act 1908 and was in breach of corporate law. The judge

Tram's off again ! Sadly not an unusual occurrence, No. 5 has derailed and crossed the road and ended up on the opposite pavement. Derailments of wagons, locomotives and passenger cars were relatively common. The jacks and sleepers on the road are about to be used to rerail No. 5. The open wagon behind the engine belongs to Clark's Flour Mill in Wantage, a major provider of goods traffic to the tramway. *John Scott-Morgan Collection*

allowed the company additional time to register the mortgage, extended the limit to 21 February 1922, which the company was able to meet. The loan had been made by the Wantage Tramway Company to John Arbery in order for him to purchase four pieces of meadow land, a garden and buildings belonging to the company which they had no use for and wished to sell off.

The ownership of the company had changed by 1924 with the Loyd Lindsay family no longer involved in the company since the death of Lord Wantage. The board of directors also changed. The majority shareholders were now all local business owners with Percy Clark, the owner of the flour mill in Mill Street the main shareholder owning 190 ordinary and 100 preference shares, and Arthur Colbert, an accountant living in Charlton near to Wantage holding 230 preference shares. Edward Ormond, who had been the company secretary had passed away, with his shareholding of 81 ordinary and 84 preference shares passing to his son Edward Brooks Ormond, who held 20 ordinary shares in the company in his own right and had been a director of the company since his father had retired in 1917. The final large shareholdings were held by Henry Saunders, a retired chemist from Wantage, who held 10 ordinary and 50 preference shares and Charles Collins, a farmer from Grove, with 40 ordinary shares.

A large number of individuals held the twenty shares which entitled them to stand for election as directors of the company, with the ownership of the company now reaching far away from Wantage. Shareholders could now be found in Newbury, London, Avebury, Barnstaple and North Shields near Newcastle upon Tyne. The directors of the company had been elected from this group, with all of them living in Wantage. Percy Clark (miller) had been elected as chairman of the company, with Edward Brooks Ormond (solicitor) as company secretary. The other directors were Henry E. Denis de Vitre and Henry Saunders (gentlemen of independent means), with the board being completed by Ernest T.W. Nalder (engineer and director of Nalder and Nalder).

1923 was a much better year for the company, although the passenger traffic continued to decline. The tramway was having to compete with a motorbus service which the GWR had started. The bus service from Wantage Road served most of the towns between the station and Swindon, including Wantage. The bus fares were considerably cheaper than the fares on the tramway, and with the GWR's bus service departing more regularly throughout the day, local people began to travel on the road service instead of the tram.

Ironically, the tramway received a deputation from the GWR staff at Wantage Road station requesting free passes for travel over the tramway. The tramway directors pointed out that there was a limit to the number of free passes which they were allowed to give out under the conditions of their authorising acts and other agreements. As such the company was only able to offer free passes to the stationmaster, the station's clerical staff and their families. Signallers and yard staff could not be offered a free pass, but the tramway was able to offer the men and their families reduced rate season tickets for an annual charge of 7s 6d. All of the offers were accepted.

The aging locomotives caused the company problems once more in March. No. 5 broke a crank pin and had to be sent to Swindon for repairs. Charles Collett, the GWR's Chief Mechanical Engineer, intervened on the tramway's behalf and produced a plan of action to quickly have the repairs completed and the engine returned to Wantage. No. 7 had developed boiler trouble and was found to be in urgent need of a retube just as No. 5 failed. Collett helped the tramway out again, quickly supplying the necessary boiler tubes, which arrived at Wantage Road and were taken straight to Nalder and Nalder's works in Challow where the Manning Wardle engine had been dismantled and the boiler quickly removed and repaired and was waiting for the tubes to arrive. The work was completed quickly and No. 7 arrived back on the line to take up goods and passenger duties once more. In the meantime, the only locomotive available to the tramway had been the Matthews

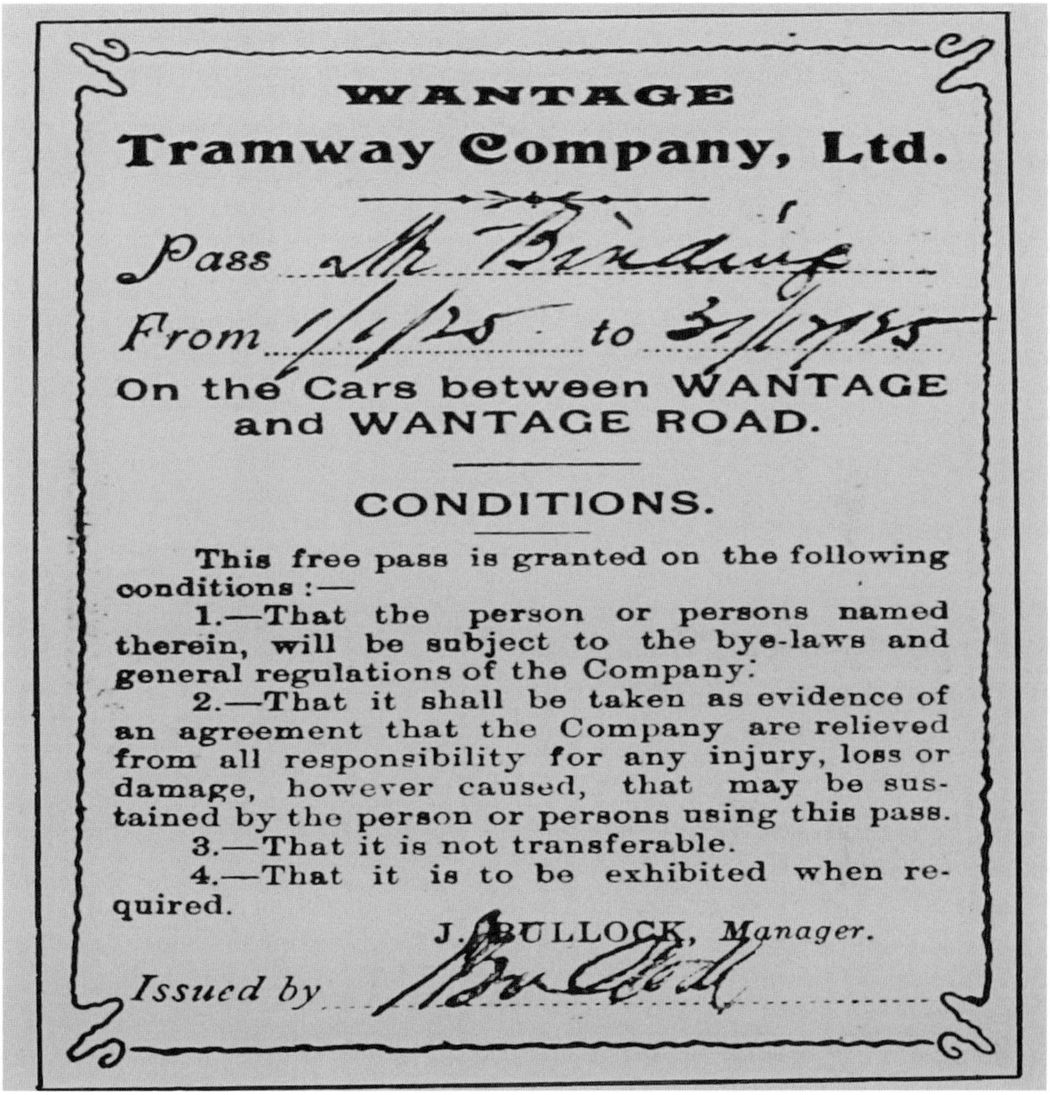

WANTAGE

Tramway Company, Ltd.

Pass ...

From ... *to* ...

On the Cars between WANTAGE and WANTAGE ROAD.

CONDITIONS.

This free pass is granted on the following conditions :—

1.—That the person or persons named therein, will be subject to the bye-laws and general regulations of the Company.

2.—That it shall be taken as evidence of an agreement that the Company are relieved from all responsibility for any injury, loss or damage, however caused, that may be sustained by the person or persons using this pass.

3.—That it is not transferable.

4.—That it is to be exhibited when required.

J. BULLOCK, *Manager.*

Issued by ...

Wantage Tramway Workman's Pass. 1920s. *Vale and Downland Museum, Wantage*

tram engine which was not very suitable for goods traffic. Nevertheless, necessity meant it had to handle some of the goods traffic, with the remainder sent by road using the company's two new road lorries.

Anglo-American approached the tramway in April to request a new lease on some additional land next to their oil storage depot which the tramway was not using. It was eventually agreed on the same terms as the lease granted earlier in the year but with an additional £7 7s per year rent with the new lease being signed only at the beginning of 1923.

No. 7 in the lower yard by the goods shed, in May 1930. The bar attached to the front buffer beam is the coupling used when the locomotive was hauling the tramway's passenger cars. *John Scott-Morgan Collection*

Bullock was working on improving the state of the tramway's line, and was authorised to purchase another consignment of used rails from the Midland and South Western Junction Railway as well as some second hand points from them. Repairs were also needed to the tramway's passenger car No. 3 which required new side panels. The tramcar was repaired and put back into traffic despite the dwindling passenger traffic on the line. The general state of the passenger cars was not good by this time, and the Ministry of Transport received a complaint about the condition of the tram cars. The Ministry asked the tramway to respond to the complaint, and the company replied stating that the two cars in question were perfectly safe and since the time of the complaint had received a thorough clean. The tramway had also requested quotes from tramcar manufacturers for the purchase of a new passenger car, although later events would see the tramway never buy its planned new tram car.

1923 had been a tough year for the company in some respects. A large amount of money had been spent on repairs to two of the locomotives during the year which eroded the company's profits considerably, although the company still returned a sufficient profit for the directors to declare a dividend against both the preference and ordinary shares after a portion of the profits had been put aside to fund permanent way repairs.

The directors informed the shareholders at the annual general meeting on 30 November 1923 that the full payment of the 4½ per cent dividend to the preference shareholders still left sufficient funds for the company to pay a 3 per cent dividend on ordinary shares. It was also agreed that the funds put aside for the planned repairs to the line should be used for that purpose, but that a local contractor should be used to undertake the work. The tender documents were duly sent out only to civil engineering contractors in and around Wantage.

1923 ended with a minor accident which reflected the state of the tramway's line. A truck had derailed in the tramway cutting and had damaged the rails as well as itself. It became clear to Bullock and the directors that the repairs to their permanent way were becoming an urgent priority for the new year.

1924 began with the company looking at more repairs on the locomotive fleet and some staffing challenges. No. 7 had been working hard and was becoming in need of an overhaul once again, but the work was delayed until March, the GWR agreeing to

A fine portrait of all three of the tramway company's operational locomotives in the early 1920s. The Matthews tram is on the left with its skirts all in position. No. 7 is in the middle of the lineup, but with its original small cab, which was later replaced by the GWR at Swindon with the larger cab it ended its operational life with. No. 5 is at the front of the line, the steel plates with which it was fitted to meet the requirements of the board of trade can clearly be seen, number 7 never being fitted with similar 'deflectors'. Note the rails laid on longitudinal sleepers with cross ties. *Vale and Downland Museum, Wantage*

undertake the work when the tramway was ready to send the engine to Swindon. Fitter White, responsible for the daily maintenance of the tramway's three locomotives was summoned to the 30 January directors' meeting to answer questions about the position he held, and the performance of his duties. The meeting was arranged just to clarify what White was responsible for and his working hours, which resolved White's concerns regarding his position in the short term. The early spring presented the company with a potential staff shortage when driver Kent informed the tramway manager that he wished to leave the company. Already lacking a driver after the dismissal of driver Clarke in 1922, advertisements were quickly placed for the vacant positions.

With goods traffic increasing again in the early part of the year, the company gave Bullock authority to recruit an additional clerk for the goods office, on a wage of 45s per week. The other side of the company's business, the passenger services, were beginning to cause the company a great deal of concern. Passenger numbers were falling dramatically as people continued to opt to use the GWR's bus service between Wantage and Wantage Road station. Despite efforts to improve the passenger cars' appearance, and slight reductions in the tramway's passenger fares, the services continued to see fewer passengers. In late March, is was decided that, in the short term, only one driver was to be recruited to replace driver Kent, the other position to remain open pending a decision on the future of the passenger services.

With the company already examining its income, a notification was received from Paddington which was unlikely to have been welcomed. The GWR wrote to inform the tramway that it was reducing the allowances it paid to the company for deliveries undertaken on its behalf from 4s 3d to 3s 9d per ton from 1 May. The planned cut in revenue from the GWR resulted in the tramway's manager looking at the outstanding debts owed by the tramway's customers. Letters were sent out reminding the people concerned that their accounts were overdue, most settling by return mail. One of the tramway's customers, Mr. Creese, a farmer, had not responded by the end of May, and the manager was authorised to send a final demand, and if Creese had not settled within a week, to begin legal proceedings. The final reminder worked, and the debt was paid off by the end of June.

Whilst passenger traffic was in decline, goods traffic continued to grow, which left the company still profitable, but looking for more income to replace the passenger fare revenue. Rental of property had been undertaken by the company since it was first formed, and in April 1924, the tramway was approached with a request to lease land along the side of the line. Post Office Telephones asked the company to allow it to place telephone cable posts along the side of the line. The directors agreed to allow the poles to be erected on its land in return for a rent of 2s 6d a year. Post Office Telephones agreed, and the poles were erected during June.

A second new tenant approached the tramway on 30 April, this time requesting use of a parcel of land in Wantage adjacent to the upper yard. The Anglo-American

Oil Company, based in Westminster, wished to take a lease which would allow them to construct corrugated iron buildings, and tanks for storing oil and motor spirit (petrol) and to lay the necessary pipelines for the tanks. They also wished to be granted permanent right of way into the yard in order to access the site of their facility and to enter into a subsidiary agreement for the conveyance of all their oil and petrol along the line in both directions. The tramway company offered Anglo-American a lease on the terms they requested for the sum of £17 7s for a period of 21 years. Carriage charges for the oil wagons were to be agreed separately, but the tramway directors offered the oil company free conveyance of empty wagons in any direction along the line. The terms were agreed and the contracts signed.

No. 7 had been sent to Swindon in late May, but maintenance work was not only focused on the locomotive fleet. The tramway line also received work during the summer. Three sets of second hand points were purchased from Swindon for £10 15s and installed in the yards at Wantage, replacing worn out trackwork. New sleepers shortly followed which were installed along the entirety of the line. The maintenance work on the line had left a large quantity of scrap in the yards at Wantage which was sold in July.

The high volume of freight traffic meant that the cranes in the upper and lower yards were getting a lot of use. The two cranes were already old when the tramway purchased them, and by 1924 both were worn out. The lifting chains were removed in late June, with one set replaced by a new chain purchased by the manager, the second set, in better condition, was sent away for repair. The repairers advised that the chain had been overloaded at some point in its working life but once repaired would be good for many years work.

The conversations that the directors had with White in January seem not to have resolved his concerns, and he left the company in early August. White's assistant Miller had applied to replace him, and in view of the high quality of his work the directors were pleased to promote him. Miller took over as chief fitter on a wage of £2 10s per week at the end of August. It was decided not to hire a new fitter's assistant at the time because of the declining passenger traffic.

Passenger numbers continued to decline as winter approached. The manager was instructed to open discussions with the GWR about their competitive new service. Bullock wrote to Paddington in late October regarding the motorbus that was running between Wantage Road and Wantage, and then on to Swindon, picking up passengers at the intermediate villages. The bus service seemed to be more economically viable, and so the tramway's manager was authorised to open discussions with the GWR about the future of the passenger tram service.

At the annual general meeting of the company held on 5 December 1924, it was announced that the company had once again had a good year, despite the competition from the GWR's bus service. As a result, the tramway would be providing a dividend

No. 5 hauling a passenger service consisting of passenger car 4 and one other unidentified car. The bar attached to the coupling hook was used for attaching and decoupling the engine to and from the passenger cars. A similar arrangement could be found on No. 7. Both of the tram engines, Nos. 4 and 6, seem to have had a similar arrangement. *John Scott-Morgan Collection*

to the shareholders for the financial year just ended. Profits had been high, and it was announced that after the payment of the 4½ per cent dividend to the preference shareholders, sufficient funds were available to not only place a good amount into the reserve fund, but to also pay a dividend of 5 per cent on ordinary shares.

The end of the year saw the tramway back to a full complement of locomotives. No. 7 had been repaired and returned from Swindon. Locomotive No. 5 was then sent to Swindon for a light overhaul, and had returned from the GWR at the start of December. The two tank engines were reported as working satisfactorily, but the Matthews tram was not sent for an overhaul in 1924, pending decisions on the future of the passenger service.

Chapter 12

THE END OF THE PASSENGER SERVICE

The new year began with the tramway directors accepting the reissued traffic rates from the GWR. The tramway's management had met with the GWR's representatives in late 1924 with the railway company agreeing to apply less of a cut to their rates. The tramway agreed to the GWR's new payments and informed the railway accordingly on 2 January.

Passenger revenue had continued to fall and in the new year the situation was becoming critical. The tram's cars often ran lightly loaded or even empty and discussions were ongoing with the GWR about their motorbus service and internally at the tramway about whether their own service could be made viable once more, or if it should be abandoned. In the climate of falling revenues and a large wage bill, the company directors were reticent to agree any pay increases at the start of 1925. Bullock approached the directors on 25 January requesting a pay increase, which the directors informed him was impossible. They did make Bullock a counter offer which he accepted. It was agreed that the manager would be paid a bonus of £10 when profits were between £700 and £799 with a further £10 for every hundred pounds the annual profits were in excess of £799. The company was still profitable and very much viable, goods traffic providing large profits, although the drain on the company's finances from the increasingly unviable passenger trams was causing the caution among the directors. The company was, however, able to donate £5 5s toward the construction of a new hospital in Wantage in January.

The company faced another financial challenge when the GWR's bill for the overhauls of the two tank engines arrived at Wantage in early February. The company directors decided not to pay the full amount of £651 7s 10d, preferring to only pay £351 7s 10d on 18 February, accompanied by an explanatory letter informing the railway that the balance of the invoice would be paid later. The company was able to pay the full amount, but preferred not to since full payment would have taken the company's bank account below £400.

A new source of revenue presented itself to the tramway when the Anglo-American Oil Company approached the general manager requesting that they be allowed to occupy additional land adjoining the tramway's yard in Wantage, where they intended to build a storage tank to hold petroleum. The oil company's manager informed Bullock that the additional petrol storage tank would guarantee more traffic for the tramway as all the fuel would be delivered in full tank loads by rail. Bullock referred the request to the directors who delayed responding since they were concerned about the fire risk and needed to seek advice.

The tramway's manager was asked to make enquiries as to the conditions under which petrol tanks may be erected at railway stations. Bullock requested the assistance of the GWR once more and received help from the railway's district goods manager at Bristol. The tramway was sent copies of the GWR's standard agreements which the railway already had in force with customers who were holding petrol on Great Western property near stations. The railway gave the tramway company permission to copy the agreements and use them for their own purposes. This was done and the Wantage Tramway Company version of the railway's legal contract was sent to the oil company. The tramway company received no response to their proposed agreement for the use of more of the company's land at Wantage, and were still waiting at the end of May.

The tramway temporarily lost a driver in March when driver Banning informed the company that he needed treatment for an injury and would be away from work for a short time, but asked that his job be kept open for him. The tramway assured him that his position was safe and wished him a speedy recovery. In order to fill the gap in the company's driver roster, the manager was asked to approach one of their old drivers and ask them if they would be prepared to return for a short time to help the tramway. One agreed, John Hewitt returning to his old engines.

Meanwhile, the GWR's chief accountant had written back to the tramway acknowledging receipt of the part payment of the maintenance invoice from Swindon and advised that the GWR was perfectly content to extend payment terms to the tramway if that would be of assistance. The railway's accountant asked the tramway to confirm when they would be likely to pay the remaining £300 balance so that they could make the necessary allowances in their books. The tramway promised the final payment would be made at the end of May but sent the cheque to the GWR early on 20 May.

At the same time that the last payment for the work done by Swindon was being made, the tramway was planning on more work on the locomotives. No. 5 was to be retubed during the autumn, and so the required boiler tubes were ordered from Swindon, with the work planned to be undertaken by Nalder and Nalder in Challow. The retube would cost less than estimated, since some of the tubes that had been removed at the last boiler overhaul had been inspected by the GWR and the company's boiler insurer, who had cleared them to be reused.

The declining passenger traffic on the tramway had continued to worsen, leaving the company no option but to suspend some of the quieter services. As a result,

No. 5 with passenger car number 4 in the upper yard in 1925 prior to the end of the passenger service. *John Scott-Morgan Collection*

fewer staff were needed and so the manager had been instructed to dispense with the services of one of the passenger conductors and one of the permanent way labourers. Both men left the tramway's employ at the end of May 1925. With passenger traffic in what appeared to be terminal decline, partly caused by the GWR's bus service, and partly by other local means of transport, the company directors arranged for two representatives to meet the GWR's general manager, Sir Felix Pole, at Paddington to discuss the Great Western's future policy in regard to bus services from their stations, especially from Wantage Road. Whilst the upheaval related to the decline in passenger traffic was ongoing, the company was pleased to see driver Banning return to light duties on the tramway on 22 June, with him expected to be able to resume his duties driving the tramway's goods engines at the beginning of July subject to his doctor certifying him as fit.

The two directors returned from their meeting with Sir Felix Pole and reported back to the rest of the board at their meeting on 10 July 1925. The GWR was intending to continue with the road bus service between Wantage Road and Swindon.

The railway bus was thought to be likely to continue to erode the passenger traffic, which was already viewed as only just covering its own costs. The directors agreed that the passenger service was not a viable long term product to be offering the public and so it was arranged that the company secretary would contact the GWR to arrange a meeting to discuss the fate of the passenger service. The meeting took place during July with Michaels, the GWR's Superintendent of the Line, and during the discussions the representatives from the tramway suggested that the GWR should take over all of the passenger and parcels traffic from Wantage Road, with the tramway then halting the passenger trams. Michaels promised to consider the proposals and seek an opinion from Paddington.

A letter was received at Wantage from Michaels shortly after the date of the meeting, in which he confirmed that the GWR would take over the operation of all passenger and parcels services from Wantage Road from 1 August if the tramway were to confirm that they wished them to do so. An extraordinary general meeting of the tramway company's shareholders was called for the proposal to be put to the investors since they would need a vote to accept the cessation of passenger tram services. The meeting was duly held and the motion was passed. The tramway was to hand over all passenger and parcels traffic to the GWR from the beginning of August.

With the passenger services to be abandoned, the company would have to be reorganised with staff redeployed or laid off. The manager's salary would be reduced to £250 per year, with the company's managing clerk Hale to also have a pay cut to £2 per week and to move to the goods depot in the lower yard. The company's other clerks, Douse and Holland, were given notice of dismissal as there would no longer be enough work for them to do. The two 'parcels lads', Rolls and Honey, were also served notice as was the goods clerk Bressell. The cessation of the passenger trams meant that the company now had too many drivers and as a result drivers Manning and Weaving were given notice of dismissal, although Manning was eventually retained by the company.

The staff cuts continued, Willis, the yardman's assistant was served notice, with the passenger tram conductor Hathwell taking on the role of yardman's assistant. The other passenger conductor, Skinner, was given the option of staying on in the permanent way department on a reduced wage of 35s per week, with Prior in the department to be laid off if Skinner wished to take up the option. The directors informed all those affected by the withdrawal of passenger trams that they 'regretted having to take such distasteful actions', and announced they would pay everyone in the company an additional week's wage. The passenger side of the tramway business was to be completely closed down, with the passenger engine, No. 6, to be laid up with the passenger cars also being stored.

The manager's salary cut was delayed until 1 September, and with Hale being taken on by the GWR and not moving to the goods depot, Holland was to be retained as goods clerk on a weekly wage of £2. On 16 August, notice was served by the tramway of their intent to cancel their contract with the post office, and the Postmaster General agreed to end the mails contract by mutual consent, the GWR shortly thereafter agreeing a new contract for carriage of the mails.

Further negotiations were undertaken between the GWR and the Wantage Tramway Company at the end of August, with the GWR agreeing to also take on the mail services that the tramway had been operating for the post office. The railway company agreed to take out a six month lease of the tramway's offices in Mill Street to act as their temporary town office at a rate of £63 per year. Further consideration would later be given by the GWR for a more permanent arrangement which would be more advantageous to the tramway company.

The railway agreed to employ Douse and Hale on the same wages as they had been paid by the tramway. The conductors, Skinner and Clark, were to be interviewed by the GWR and taken on as conductors on the bus service if the railway company thought they were suitable. The GWR also looked to employ as many of the other displaced tramway staff as they could at Wantage Road, Didcot and Uffington Junction stations.

Among the negativity associated with the cessation of the tramway's passenger services, a slight ray of sunshine arrived in the offices at Wantage. The Anglo-American Oil Company finally agreed to the terms offered to them by the tramway for the lease of tramway land for the storage of petrol in new tanks and an increase in fuel tanker traffic on the line. The final draft agreement for the land was provided to the oil company on 30 September, and after a few amendments was signed by both parties.

The end of September saw the tramway company reviewing the equipment that was in the two yards at Wantage with a view to disposing of anything no longer needed. The company had laid up the Matthews Tram engine and the three remaining passenger tram cars, none of which would run again. The manager was instructed to dispose of them, but it proved impossible to sell them on and they remained in the yard at Wantage, lying derelict, until the tramway was finally closed in 1946. The GWR had a small immobile petrol tank in the annex to the tramway's locomotive shed which held the equivalent of thirty tins of petrol. The GWR was asked to remove it as it was now against the company's fire insurance.

Goods traffic had risen to record levels during 1925. Goods trains were regularly made up of up to twelve wagons mainly carrying coal for the gas works in the town, agricultural merchandise and parcels. The main freight leaving Wantage was still flour from Clark's mill, despite the earlier cancellation of the annual freight agreement between the two companies. A large proportion of Clark's flour was sent direct to Huntley and Palmer's factory in Reading which also had a direct rail link. The direct rail connections were of great benefit since flour dust in the air can

Matthews Tram Engine, Wantage Tramway No. 6 derelict by the orchard in the lower yard accompanied by passenger cars 5 and 4. 1930. *John Scott-Morgan Collection*

potentially be explosive if exposed to an ignition source. The bulk transport of flour from mill to biscuit factory was also beneficial in reducing the cost of the ingredients Huntley and Palmer were using.

The goods traffic only needed two locomotives, Nos. 5 and 7 proving more than adequate in handling the work. The personal journal of J. Bullock, clarifies the fate of the remaining locomotives which had been used on the tramway in an entry for February 1926. Locomotive No. 4, the Hughes Tram, had been scrapped at Wantage as it was worn out and not worth repairing since there was no longer any work for it long before the cessation of the passenger service. Locomotive No. 1329, referred to in the journal as 'the South Devon Railway Loco' had also been scrapped. The Matthews tram remained intact, although relatively derelict, in Wantage Town yard with no decision having yet been made on its future. Bullock's journal confirms that the locomotive was the 'James Matthews Tram which was trialled on Liverpool Corporation Tramways routes in the city immediately after it had been built'.

The mortgage owed by John Arbery had been paid off in full on 17 July 1925, with the company relinquishing any financial interest in the land.

The closure of the Wantage Tramway to passengers did not receive a great deal of public attention. The cessation of the passenger service received only a brief mention in the *Locomotive Magazine*'s 15 January 1926 edition.

An article in the 9 January 1943 issue of *Modern Transport Magazine* described the passenger service as 'not exactly speedy' but convenient. The tramway was still picking up and setting down passengers outside their homes if they were alongside the line. It seems that various methods of stopping the tram on the line had been agreed between the drivers and conductors by the people who lived along the line, and whilst not fast, the friendly service probably offset the cost of the tickets in the minds of the regular passengers.

The carriages were quite luxurious despite only having a single class of travel after the First World War. The clerestories of the carriages were fitted with stained glass as were the fixed lights above the passenger compartment windows. The carriages were all fitted with plush carpets and comfortable seats with cushions.

The luxury of the carriages was not enough to save the passenger service. The number of passengers had declined hugely by the mid-1920s, still mainly impacted by the motor bus which was being run by the GWR between Swindon and Wantage Road station which continued to call in the town on its route. The bus was both quicker and cheaper than the tramway and most passengers now preferred the railway company's road service over the tram.

Goods traffic had remained profitable though, with a lot of freight being carried along the line from and to Wantage Road. The flour produced in Clark's mill in Mill Street was all being carried out of Wantage on the tramway, as were coal and other materials that were being delivered to the mill.

Despite the steadily increasing goods traffic the tramway company was not using the old passenger station line and run round loop at Wantage Road, and it had become derelict. The GWR, needing to expand its facilities at Wantage Road to accommodate the increasing goods traffic, approached the tramway with a suggestion that they should, at the tramway company's expense, demolish the old passenger terminal at Wantage Road, and also lift the related line and loop. The railway wished to use the land to construct additional siding space and a better road to access the goods yard.

The tramway company agreed and a new agreement was drawn up to that effect as a supplemental part of the original 1875 working agreement and the 1912 supplement which covered the new coal yard at Wantage Road which the tramway had wished to build and that was still in use by the same coal merchant.

The 1925 financial year ended well for the company. The company had been profitable again with a surplus of £329 13s 5d generated despite the closure of the passenger service. The directors announced the healthy financial position of the tramway in their annual report. When the report was presented to the shareholders at the annual general meeting held on 10 December 1925, the chairman announced that after the 4½ per cent dividend on preference shares, sufficient funds remained to pay a dividend of 4 per cent on ordinary shares.

As 1925 ended, driver Manning was off work with long term sickness and the tramway's fitter Miller had been driving the goods trains in his absence. Miller had to stop on 24 October when he got red hot ash in his eye and was ordered to stop driving by his doctor. The company approached their old driver, Weaving, who had been laid off in July, asking him to return which he gladly accepted.

The GWR relinquished occupation of the tramway's old offices on 30 December following the end of the six month agreement, and the refusal on the part of the tramway to allow the railway to remain in the building on half the rent which had been agreed for the initial lease period. The directors decided to keep the building but to lease it out and signs were erected to that effect. An offer was finally received from a Mr. Parry offering £710 for the Matthews engine and the three redundant passenger cars, but the offer was viewed as insufficient and refused.

The premises in Mill Street remained unoccupied at the end of January 1926, no applications to lease the property being made to the company. The directors decided to

A view of the derelict passenger cars stored in the lower yard at Wantage in September 1930. They would never be used again. *John Scott-Morgan Collection*

A close up study of the derelict Matthews tram at Wantage during May 1930. The side skirt has been left in the open position giving a good view of the arrangement of the wheels and motion and particularly the long piston rod. The locomotive appears to be relatively complete other than the removal of all bar one of the side windows. *John Scott-Morgan Collection*

dispose of the former passenger offices and the manager was instructed to erect a sign on the outside of the building to that effect. The company needed to gain the agreement of the shareholders before the offices could be sold and an extraordinary general meeting was arranged at which it was agreed that part of the building at 14 Mill Street should be sold. The passenger offices, tram shed and platform were to be sold off. The house and shop which the company owned at 12 Mill Street were to be kept and leased to a new occupant who had applied to take on the premises for their business.

Goods traffic was increasing once again in the first quarter of 1926. It was reported at the 28 April directors' meeting that the company's freight services were increasingly busy. In just the eight days prior to the meeting, the company had exchanged around 300 trucks of tarmac and stone with the GWR. Other goods traffic was similarly increasing and the cash on hand at the bank was rising.

Despite the high volume of traffic that the tramway was carrying, part of their facilities at Wantage Road renamed unused. In late May, the company received a letter from the GWR's divisional engineer at Bristol, Mr. Quartermaine, about the yard. Quartermaine noted that the tramway's old passenger terminal on the GWR's property at Wantage Road was out of use and the land would be of great use to the railway. If the tramway had no objection, Quartermaine wrote, the GWR would prefer to remove the two old passenger lines. The tramway replied that they did have an objection. The company had paid a large sum of money to the GWR for the sidings to be laid in the yard, and that they may become useful to the tramway for laying aside a few wagons if 'a great presence of traffic came along'. The sidings remained in place after a meeting with the engineer at which their retention was agreed, although they were to be shortened and the old run round loop was to be disconnected at the station end and made into a dead end.

The summer passed with little incident, although some traders had not been paying their carriage charges due to the company. In August the debt position had become untenable, and the tramway began the process of asking for overdue accounts to be settled. Most were paid during August, but it was found that the GWR had not paid any of the rent due for the six months the railway had occupied the tramway offices in Mill Street. The GWR was approached for payment of the outstanding account.

A tenant was found for part of the old passenger premises, and as a result the company decided to retain all of the buildings. The Cement Marketing Company had approached the tramway wishing to rent the shed adjoining the locomotive repair shed as a cement store. The cement was to be delivered to the store by the tramway, who would also be contracted to carry the outbound traffic to the cement company's customers. The tramway agreed, and the contract was signed.

The end of the financial year for 1926 saw the company's profits decline a little from the previous year, but nonetheless the company returned a good profit on its activities. The company was able to pay a dividend of 2 per cent to holders of ordinary shares after the payment of the statutory 4½ per cent dividend upon preference shares. The board of directors was changed at the year-end annual general meeting with Percy Clark elected chairman and Henry Saunders replacing Ormond as the company secretary since the latter was often away from Wantage for many weeks and had asked to stand down.

1927 and 1928 proved to be relatively quiet years for the company when compared to the upheavals of the previous few years. The traffic continued to grow dramatically with a large profit generated in both years. The old passenger arrival platform at Wantage found a new use when the company decided to convert the derelict building to garages for lease to local people during summer of 1927. The tramway offices were not sold, and remained in the ownership of the company, with a new tenant taking occupation in June the same year. The Ministry of Works moved in at the end of the month.

The GWR and the tramway came to an agreement regarding the retention of the sidings in the yard at Wantage Road. The tramway company agreed to pay the railway £9 per year for the cost of the upkeep and maintenance of the lines and sidings.

A very derelict passenger car dumped in the old approach to the passenger terminal train shed. It seems that the gauge of the rails has spread and the car has fallen between them at some point. The corrugated wall behind the tramcar is the end of the old train shed which had been boxed in to convert the building to motor car garages. *John Scott-Morgan Collection*

The tramway had had a very successful two years, the profits for 1926-1927 far in excess of the earnings for the 1925-1926 financial year. The company directors were able to announce a good dividend of 6 per cent on ordinary shares after the 4½ per cent dividend required on preference shares, as well as adding a considerable increase to the cash in the reserve fund.

1928 was also good for the tramway. Profits had remained high, and the company was able to add to its reserve fund as well as announcing, at the 14 December 1928 annual general meeting, a 4 per cent dividend on ordinary shares, after the 4½ per cent payable to holders of preference shares.

The company decided that, despite returns being good and traffic continuing the grow, the workload no longer justified the position of a general manager. Joseph Bullock was given three months' notice of dismissal on 29 December 1928.

The decade ended with the company remaining highly profitable. The directors were again able to add to the company's reserve fund, and after paying the 4½ per cent dividend to preference shareholders, were able to announce a 5 per cent dividend on ordinary shares at the 23 November annual general meeting.

Chapter 13

THE 1930s AND INTO WORLD WAR TWO

The new decade began with a change at the helm, the board of directors being reduced to just three men. Percy Clark remained the company's chairman, with Edward Brooks Ormond returning as company secretary, the third director was Henry Saunders who had also decided to stay on. The other directors had stood down, the remaining three men seeing the company through the remainder of its existence. After the changes at the helm the 1930s were relatively quiet for the company other than the usual regular derailments of trains on the line.

The first financial year of the new decade resulted in the company remaining profitable, the shareholders once again receiving a return on their investment.

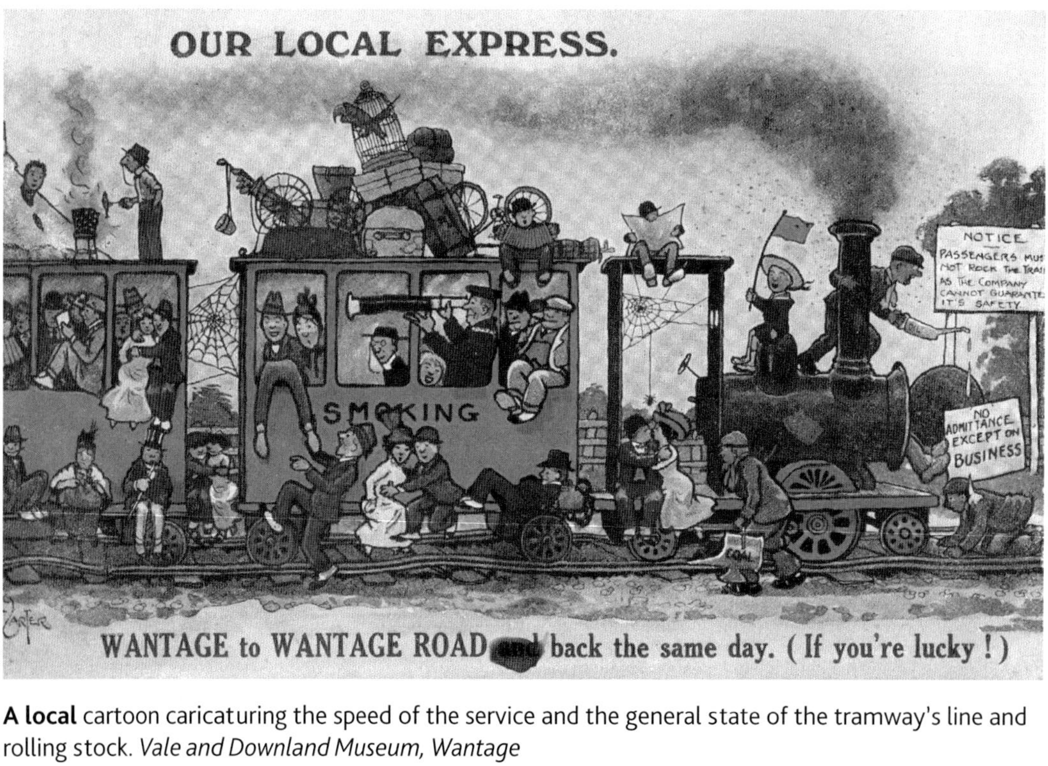

A local cartoon caricaturing the speed of the service and the general state of the tramway's line and rolling stock. *Vale and Downland Museum, Wantage*

The group attending the 6 December 1930 annual general meeting were informed that the 4½ per cent dividend on preference shares would be paid, but holders of ordinary shares would receive a 5 per cent dividend. The company remained profitable throughout the first half of the 1930s, the preference shareholders receiving their 4½ per cent dividend and the ordinary shareholders a 5 per cent dividend every year until 1934. The profits were so high for these years that the directors were able to transfer money to the reserve fund to cover repairs and maintenance.

The company ended the 1935 financial year in a good position. Profits were slightly reduced but still good, allowing the directors to declare, again, a dividend to the tramway's shareholders. The preference shareholders once more received their full dividend, with ordinary shareholders being granted a 4 per cent return.

The *Railway* magazine reported in 1936 that the tramway's main traffic by then was principally coal for the town with regular large loads also being delivered to the town's gas works. The byproducts, coal tar, coke and ammoniacal liquor were removed by the tramway for onward despatch on the GWR. The tramway was carrying regular consignments of petrol for delivery to the Anglo-American Oil company's storage facility on the edge Wantage Town station yard. It was noted that by 1936 only locomotives Nos. 5 and 7 were in traffic. No. 5 was said to have gained a steam brake by the time the article was written, although No. 7 had not been provided with such a convenience.

Wantage yard had become something of a rolling stock graveyard by the 1930s. No. 6 had run until the withdrawal of the passenger service since which it had stood derelict on an overgrown siding with the passenger cars. The locomotive was finally broken up by the tramway company in the early 1930s.

1936 ended with the company remaining profitable, allowing the company to pay the preference share dividend as well as a 4 per cent dividend to ordinary shareholders.

The company returned a high profit for the 1937 financial year, allowing the directors to transfer £250 to the reserve fund in anticipation of higher maintenance costs during 1938. The company was still able, even after putting aside the funds for repairs, to pay dividends for the year. The dividend on the preference shares was paid, leaving funds to pay a 4 per cent dividend on ordinary shares.

The company continued to be well run and profitable in 1938. The shareholders were informed that the company would once more be paying dividends from the profits made up until the end of the financial year on 30 September. The dividends were again 4½ per cent on preference shares with a dividend declared at 4 per cent on ordinary shares.

The clouds of war were gathering once more during the late 1930s. On 3 September 1939, the British Prime Minister, Neville Chamberlain, announced in a radio broadcast to the nation that the German government had not agreed to withdraw their troops

Another view of the Matthews tram and passenger car No. 5 lying abandoned on the overgrown siding at Wantage in 1930. *John Scott-Morgan Collection*

from Poland, and therefore a state of war existed between the two countries. The Second World War had begun.

The tramway company remained a successful business throughout the war and continued to run goods services between Wantage Road and town for all but a few months of the conflict. It remained profitable throughout the war. The company was able to pay dividends each year between 1939 and 1942 of 4½ per cent on preference and 4 per cent on ordinary shares. The trading conditions in 1943 were tougher, with a great deal of expense required on maintenance, which resulted in the company paying only a dividend of 4½ per cent on preference shares. Dividends were once again paid to all shareholder in 1944 and 1945 of 4½ per cent on preference and 4 per cent on ordinary shares. The tramway, despite its challenges, remained successful throughout the six years impacted by the Second World War.

The tramway remained very busy during the conflict with the goods services remaining extremely busy throughout. In a December 1945 article for *Railway World*, John Pope reported that the goods services on the Wantage Tramway had been well used. The tramway had been running six trips each way every day throughout most of the war, with revenues being extremely good. Each of the trains was made up of a mixture of open, flat and box van wagons, all of which were full of materials and goods.

No. 7 running light engine near Wantage lower yard in June 1939. Once the tramway left the Besselsleigh Road it looked much like any other light railway. *John Scott-Morgan Collection*

Despite the heavy usage of their goods services, the tramway was seeing its margins on transportation costs being eroded by increased costs of fuel and rapid increase in wages for its staff which were forced upon the company by the National Wages Board pay increase awards. Pay had effectively been nationalised under war time emergency powers, with all wages and salaries decided centrally by a government pay board to control the wages being paid by companies, and to prevent poaching of staff by companies through offers of high pay which would cause staff shortages and impact productivity in essential industries. Despite these challenges, the company remained profitable and able to pay dividends throughout the war years, although only to preference shareholders for the year ending 30 September 1943.

In order to reduce its overheads, locomotive No. 5 had been relegated to the role of spare engine during the war, with No. 7 working most of the goods trains on the tramway. Both locomotives had been sent to Swindon for heavy overhauls which were completed in 1941. The workshops had added GWR registration plates to the locomotives whilst they were in Swindon, with No. 5 allocated plate 209 and No. 7 receiving 210. The expense of the desperately needed overhauls would later contribute to the financial woes of the company post war, but the main problem was caused by the United States Army Air Force.

Wantage Tramway No. 5 was registered by the GWR in 1941 whilst undergoing a major overhaul. The plate seen in the image was fitted when the overhaul was completed. No. 7 received an identical plate but with the number 210. *John Scott-Morgan Collection*

The opening of the RAF airfield at Grove in 1942, construction of which began in 1941, provided the tramway with more traffic in the short term. Initially used by the RAF for glider pilot training and as a satellite airfield for No.15 Operational Training Unit of Bomber Command based at RAF Harwell, the airfield was handed over to the United States Eighth Army Air Force (USAAF) in 1943. The airfield was used as a repair depot overhauling Douglas A-20 Havoc medium bombers and Northrop P-61 Black Widow night fighters. The 31st Transport Group with three squadrons of Douglas C-47 Skytrain transport aircraft arrived to operate out of the airfield alongside the repair facilities. The tramway carried building materials for the construction of the airfield by the Royal Air Force and for the expansion of the base's facilities when the United States Army Air Force took over the airfield. One of the additions made to the base by the American air force was the construction of a collection of buildings to house service personnel outside of the airfield. The new facilities were reserved, to the reported surprise of local people, for African American air force personnel who were not allowed to share accommodation with other servicemen on the base and were to be housed away from it in the new segregated barrack facilities.

Once the base had been declared operational by the USAAF, the Americans were relatively self-sufficient, with their own fleet of heavy GMC 2½ ton, known colloquially as 'Deuce and a Half', trucks. The USAAF and RAF did not build a spur from the tramway into the base, as had been done on the nearby Lambourn branch line from Welford Park station when the airfield opened during 1943. The USAAF completely took over the station at Grove from the RAF in October 1943 during the build-up for Operation Overlord, the allied invasion of Europe, with American paratroop transport aircraft and glider squadrons moving in.

At Grove the American air force saw no need for a branch from the tramway. The United States army, of which the USAAF was still part until after the Second World War, was moving supplies, equipment and fuel on the GWR as far as Wantage Road by rail with transfer to their own trucks occurring in the yard at the station. The loaded trucks then travelled the short distance from Wantage Road to the gates of RAF Grove along the Besselsleigh Road. Here lay the beginning of a major threat to the future of the tramway.

The United States army trucks had damaged the line and left large quantities of mud across the rails. The mud was eventually cleared up by the tramway's employees, but the line had to be closed for a time due to damage to the track bed. The damage to the track was caused by a combination of two factors. The first was a result of decisions made in 1874 and 1875, when the line was being designed and built. The tramway was to run directly alongside the road, which was normal for a line of the type the Wantage Tramway Company was seeking to construct. In ordinary conditions the line would have been less likely to have been damaged, but the decision not to place stone slabs between and either side of the rails as was usual in urban tramways, and commented upon by Colonel Hutchinson in his first inspection report, resulted in a weakness in the structure of the tramway.

The line was built, as outlined earlier, not with stone slabs except on the level crossing at Grove, but with chalk and stone chippings covered with a layer of macadam. The decision was of no particular consequence in the 1870s, road traffic being light, mainly small carts or horse drawn carriages and omnibuses. The permanent way would have easily withstood the weight of these light vehicles if they happened to run down the tramway rather than along the road. The infrastructure was not designed to support the weight of a fully loaded 2.5 ton army truck, let alone regular convoys of them. The heavy trucks driving down the tramway damaged the underlying rails and sleepers.

The second factor relates to the United States Army. The Americans were used to driving on the right side of the road, rather than the left as in Britain. The macadam surface of the tramway looking like a part of the main public highway, combined with a tendency of newly arrived American army truck drivers to drive on the opposite side of the road, meant the Air Force trucks often ended up driving along the tramway rather than the road. The damage caused by repeated passage of heavily laden army trucks made the tramway unsafe for use and a temporary closure was necessary.

Wantage Tramway No. 7 with a goods train next to the Besselsleigh Road. The tramway noticeably runs directly adjacent to the main road. No fencing was required since the line operated under a tramway order although without it the line was vulnerable to being run over by road traffic. *John Scott-Morgan Collection*

As a result of the three month closure to rail traffic forced upon the tramway by the damage to the rails caused by the American trucks, the company had been unable to carry goods from Wantage Road. The company did not cease running goods services despite the forced closure of the line. The company transhipped all of the goods onto their own road lorries and continued to transport freight from Wantage Road during the cessation of rail operations. In his article for *Railway World*, John Pope had stated that the tramway had been wound up in 1945, with other reporters suggesting that the tramway had closed during the war, neither of which is correct. The tramway struggled on into 1946 before the directors gave up hope of resurrecting the business after the war. The rationale and causes for the closure of the tramway will be discussed in the next chapter.

As a result of the damage caused by the American truck drivers, the company directors applied to the government for help. The company appealed to the Ministry of War Transport in Reading for assistance early in March, to which the tramway received no answer. Writing again on 23 March the company requested a quick answer to their original letter. Still having received no response from Ministry's offices in Reading, the company wrote to them again on 16 May 1944. In the letter, the company secretary, Edward Brooks Ormond, outlined the problems that the tramway was experiencing in great detail.

The letter stated that both the company's road lorries were now extremely old, with one having just been reequipped with a reconditioned engine to keep it running. The temporary closure of the line due to the damage caused by the USAAF had meant that the company had relied upon the two old lorries to keep the operation working and goods flowing between Wantage and the rail head at Wantage Road. One locomotive had just returned from Swindon where it had undergone repairs and the company expected the GWR's account for the work to be in the region of £500. The other locomotive, currently out of use awaiting repair at Wantage Town, would also have to be sent to Swindon for overhaul. The work on both the locomotives, it was stated, would consume the entire repair reserves of the company.

Ormond also stated that the tramway company had done everything it could to control costs to keep the line open. He wrote that no salary was taken from the tramway for his work as company secretary, whilst the chairman, who was acting as the effective operations manager of the company, only took £100 per year for his long hours of work on the seven days a week that the tramway had been operating throughout the war so far. As such, Ormond wrote, it seemed reasonable to the directors to request authority to raise the carriage charges for coal and coke only in order to allow the company to continue operating the tramway. No response was forthcoming from the ministry's offices in Reading.

The Chairman of the tramway, Percy Clark, then wrote to the Ministry of War Transport's Charges Division in Berkeley Square House in London on the 9 March 1944 to try to get a resolution to the request. In the letter, Clark informed the Ministry that the company had been in communication with the Ministry of Transport in Reading. They had informed the Ministry that as a result of the three month closure forced on the company by the carelessness of the United States Army's drivers, the company had lost profits because of the need to use its road vehicles to carry on its business. The repairs that had been made to make the line fit for use had caused additional expenses which the tramway company now needed to cover.

In order to continue running the company, according to Clark's letter, the company would require increased revenues to make up the financial shortfall, but the company was already charging the maximum goods tariffs allowed under their authorising Act of Parliament and BoT order. It was pointed out to the minister that the permitted rates had not been increased since the granting of the original 1874 Provisional Order, and that after seventy years the directors felt that it was reasonable to ask for the Ministry's permission to increase the tariff charges for the transport of coal and coke to 3s 6d per ton, or even higher if wages were to increase further. The tramway had followed the pay awards made by the Wages Board, which had resulted in its wages having reached a total of £700 for 1944.

Clark also stated in his letter that it had been necessary for the tramway to 'send one engine to Swindon' for urgently needed repairs. It is likely that the engine referred to was No. 7, which was mechanically worn out by 1944. The company, according to Clark's letter, were expecting an invoice from the GWR of around £500 to cover the repair work carried out in their works. He also said that the company would need to send 'another engine', likely to be referring to No. 5, for repairs at Swindon in the very near future, as soon as the tramway company could afford the expense.

Whilst the GWR had granted the Wantage Tramway Company a concession and had agreed to increase its agency payments for transport of goods on behalf of the railway amounting to around £100 per year, Clark suggested that this was insufficient to keep the tramway viable in the short term unless the minister agreed to allow the tramway to increase its charges for coal and coke. The Minister replied, without giving permission for any increases in charges, with a letter applying pressure on the tramway to resume its services at the earliest possible instance.

The concessions referred to by Clark had been granted when the company approached the GWR for help. On 15 March 1944, the railway company had agreed to assist the tramway as much as it was able. The railway increased the allowances

No. 7 approaching Wantage with a goods train in April 1944. *John Scott-Morgan Collection*

it was paying the tramway for heavy goods traffic from 4s 3d to 5s per ton. The payments for light goods were also increased, from 3d to 4½d per ton. These allowances were backdated to 1 January 1941. The company received a specific increase in payments from the GWR related to grain traffic. The railway agreed to pay an increased allowance for all 'delivered' grain traffic from Bristol which was carried over the tramway for Clark's Mill and other customers. The allowances were increased from 2s to 2s 5d per ton for traffic delivered to Clarks, with all other grain traffic attracting an increase from 2s 3d to 2s 9d.

The tramway had, late in 1944, found funds to repair the tramway and restarted running rail services for goods between Wantage Road and Wantage lower yard once more. The rail traffic before the temporary closure of the line, and the road traffic during the period when the tramway could not run goods trains had generated sufficient revenue for the company to remain viable, and to pay dividends. The directors, in their annual report to the shareholders presented at the 21 December 1944 meeting, stated that the results for the year were very satisfactory, given that the line had been out of use for three months due to the damage done to the line by the United States Army's trucks. The shareholders were informed that the company had made a claim, as they were entitled to do under wartime laws, against the United States Army for compensation for the damage caused by the trucks. The chairman informed all present that they were afraid that the claim would not be met, the American army having refused to accept any liability despite clearly being in the wrong.

No. 5 with a short goods train midway between Wantage Road station and Wantage. *John Scott-Morgan Collection*

Trade had been good during 1945, with much optimism about the future as the war drew to an end. The war ended in Europe on 8 May 1945 with the unconditional surrender of the armed forces of Germany, although the war in the Pacific continued until the official surrender of the Japanese military was signed aboard the battleship USS *Missouri* on 2 September of the same year. The tramway had recovered from the damage caused by the United States Air Force and had run goods services throughout the year. With the war ending, the directors and staff of the company expected the traffic on the line would remain strong, although some threats were emerging and the company had decisions to make regarding its future.

The shareholders were informed at the annual general meeting on 17 December 1945 that the line was in a poor state and needed major repairs along the whole length of the tramway. The two locomotives needed heavy overhauls; repairs having been postponed for as long as possible during the war. No. 7 was working the traffic on the line alone, with No. 5 laid aside in a partly dismantled state at Wantage. The likely expenses were more than the company could fund at that time.

The goods traffic had been threatened by the sale of Wantage Gas Works to the Oxford Gas Company. The Oxford company had informed the tramway that they intended to cease the transport of coal to the gas works from Wantage Road, and in the longer term the gas works was to close with town gas to be piped in to Wantage from Oxford. The loss of the coal traffic would cause a major shortfall in the finances of the company. The directors were therefore considering the future of the company and once they had completed their deliberations would call the shareholders together in the new year.

The company had remained profitable during the 1944-45 financial year with the surplus funds enough to pay dividends to all shareholders. In retrospect, it must be wondered why the directors had authorised the payment of dividends if the tramway was expecting high repair costs in the next financial year. It might be speculated that the tramway board had already come to a conclusion about the viability of the company in the long term, and the decision to pay dividends, in that context, is understandable. The shareholders were informed that the company would pay the required 4½ per cent return on the preference shares, and a 4 per cent dividend on ordinary shares. The company proceeded to pay what would be its last trading year dividend.

Chapter 14

CLOSURE AND LEGACY

1946 would be the last year of the Wantage Tramway Company. A decision had been reached by the three directors of the company, one which they believed that they no option but to make. The shareholders were called to the Extraordinary General Meeting promised at what would be the final Annual General Meeting of the company and which was scheduled for 29 January. The shareholders were sent a circular before the meeting, on 18 January, which outlined the position the company was in.

The directors wrote that the future of the Tramway Company's business under post-war conditions had, for some time been a cause for concern. In order to continue the business, a large amount would need to be spent on repairs to the line. A new locomotive would be required, or at least major repairs would be needed to the two locomotives still on the tramway. The likely cost of purchasing a new engine was estimated to be very high, whilst the last repairs undertaken on one of the tramway's existing locomotives had cost £729. The directors thought that the cost of the repairs needed to return both Nos. 5 and 7 to traffic, which they believed could only be undertaken by the GWR, would be significantly higher. The board were of the opinion that whilst these expenses could be overcome, other circumstances had arisen which meant that the company would not be able to continue to be a viable business.

The shareholders were informed of the situation regarding the future of the line's coal traffic given the sale of Wantage Gas Works and its impending closure. The Great Western Railway had also just inspected No. 7 and told the directors that it was no longer safe for the engine to be used until it had been repaired. The board, according to the circular letter, had approached the GWR to ask if they would be interested in purchasing the tramway as a whole. The railway company answered in the negative, and the tramway board believed that in their opinion it would be a good time for the delivery services currently being undertaken by the tramway to be incorporated into the 'Great Western Railway's delivery service'. The directors believed therefore that the GWR should be asked to take on all the commercial services associated with the tramway company, as far as they were able, and once this had been done, the tramway company's assets should be realised, its small debts discharged and the company wound up.

The meeting was attended by the majority of the shareholders, all three directors and Henry Saunders, in his roles both as company secretary and as the tramway company's solicitor. Saunders opened the meeting by informing all present that he had made great efforts in 'preparation of the claim which was made against the U.S.A. forces for the treatment of the track which unfortunately and seemingly most unfairly had been refused'. Saunders stated that despite his efforts throughout 1945 there was nothing else that could be done to recoup the cost of the damage done to the tramway by the American trucks. The solicitor had not charged the tramway for the work.

The meeting was then reminded of the content of the circular which had been sent out on 18 January and informed of the rationale behind the statements made. The directors, it was said, were unanimous in deciding that it was in the best interests of the shareholders that the company's business should be discontinued as soon as possible and the tramway voluntarily wound up.

The finances of the company were currently only sufficient to pay the preference shareholders in full as was legally required if funds allowed. The shareholders were informed that those of them who held ordinary shares could still expect a payment of half the value of their shares, depending upon the income received from the sale of the company's property.

The meeting was asked to vote on three proposals. Firstly, that the Wantage Tramway Company should cease trading. Secondly, that all assets of the company should be realised, with the income used to settle the company's small debts and pay the shareholders a final settlement of their shareholdings. The last motion was for a formal meeting to be called, at a date to be decided, in order to liquidate the company.

All three motions were passed, those shareholders who had been unable to attend were informed of the result of the meeting, and the process of closing down the tramway had begun.

In order to wind up the company, an auctioneer needed to be appointed to handle the disposal of the company's assets. The liquidators appointed Henry Butcher and Company of Chancery Lane in London to represent the tramway company's interests as the line was closed down.

Once the directors had decided that the tramway had no future, they entered into discussion with the GWR with a view to ending their agency agreement with the railway. A letter was sent to the railway company's Chief Goods Manager at Paddington explaining the situation to the railway. The letter, dated 29 January 1946, stated that, following a meeting with the shareholders, it had been decided to discontinue the activities of the company as soon as possible. The tramway directors stated that their 'engines have not been on the track' since 21 December 1945 and would not do so again until they came down to Wantage Road to be disposed of. As a result, the directors stated that the agreements between them and the GWR would no longer be able to be operated and requested termination forthwith.

An internal GWR memorandum dated 9 May 1946 from the Great Western's chief goods manager's office outlined the agreement which had been reached between the railway and the tramway companies. The agency agreement dated 1 October 1875, with the supplemental agreements dated 13 August 1912 and 9 March 1928 were to be terminated by mutual agreement from 31 March 1946. The GWR had also agreed to purchase the junction and 88 'lineal yards' of contiguous siding located on GWR property at Wantage Road, all of which were now regarded as part of the Great Western's facilities there. Initially offering the sum of £25 for these assets to the auctioneer appointed to dispose of the assets of the tramway, the company had agreed to raise their offer to £50 in view of the long term amicable relationship the GWR had with the Wantage Tramway Company. The tramway company accepted the GWR's offer and the sale was completed on 11 May 1946.

The disposal of the company's assets continued. The freehold of the lower yard, and the coal yard occupied by Weedon's coal merchants' business in the upper yard were sold to B. Smith, whilst the freehold of the majority of the upper yard was sold to J. Waller and George Chandler. The remainder of the upper yard was sold separately, A.W. Austin purchasing the block of garages which had formerly been the passenger station, and H.C. Moise buying the former passenger office and waiting room. The other buildings and remaining equipment belonging to the tramway company, including the two tank engines, were to be sold by auction.

Once the directors had decided to wind up the company and had arranged for the GWR to cancel the traffic agreements which had been in place, taking on responsibility for delivery of goods into the town from Wantage Road station, they could begin the process of closing down the company.

The first task confronting the tramway company directors was to lodge a formal notice of intent with the company registrar. The company sent a notice of Members' Voluntary Winding Up accompanied by a Notice of Solvency, which stated that the company is solvent and capable of paying its debts but that the shareholders had made the decision to cease the activities of the company, and to liquidate it as soon as possible. The notice was submitted to the company registrar on 30 October 1946.

An extraordinary general meeting was arranged for 26 November 1946 at which a resolution was passed that the company would be wound up voluntarily. Percy Clark and Henry Saunders were appointed as liquidators of the company at the meeting for the purpose of formally closing the business down, finalising the last accounts of the company, disposing of the company's assets and paying off the shareholders.

The assets of the company that remained were to be sold off, with both No. 5 and No. 7 being part of the sale. No. 5 was the subject of a vociferous local campaign to preserve the locomotive.

R.L.P Jowitt wrote to *The Times* newspaper to explain the potential fate that awaited No. 5. The letter was published in the newspaper and gained a lot of attention. In his letter, Jowitt wrote that an old locomotive which had been the oldest in service was due to be scrapped in the near future. He also wrote that attempts had been made to donate the locomotive, which was an important example of British locomotive manufacture, to the Science Museum in London, and to the London and North Eastern Railway's museum in York. Both organisations had refused to take No. 5 since they were 'full to capacity'. Jowitt appealed to the 'good folk of Wantage' to preserve the engine. A series of letters followed on the subject of the preservation of the locomotive. Meanwhile the disposal of the company's assets continued.

The auctioneers handling the disposal of the tramway's assets on behalf of the company sent a progress report to Clark and Saunders on 2 May 1946. R.F. Howard's report informed the liquidators that the auctioneers had received the cheque from A.R. Adams for £205 for the items they had purchased, with a cheque from the GWR also received for the balance of the purchase price for No. 5. The locomotive was now the property of the Great Western and was due to be removed from Wantage Road by rail for the railway company to take it to Swindon for a cosmetic restoration. The dismantling of the tramway was due to begin shortly, George Cohen and Sons

The end, but not quite. No. 5 is being towed out of steam to Wantage Road by No. 7 where the George England tank engine will be handed over to its new owners, the Great Western Railway. No. 7 would stay on the tramway for a while longer working demolition trains before being sold. *John Scott-Morgan Collection*

having been given permission to start work despite their payment for the scrap value of the line having not yet been received. The auctioneers often worked this way, with Cohen's sending the payment immediately after their next board meeting where cheques were signed. Howard assured Clark and Saunders that Cohen's were reliable and had worked with Henry Butcher and Company on many occasions. The payment would be received in due course. Cohen's had requested that they be allowed to use No. 7 to haul the demolition trains, which was agreed.

In a letter to *The Times*, printed in the 6 May 1946 edition, R.L.P. Jowitt, whose original letter to the newspaper had drawn attention to the fate of the engine and the likelihood that it would be scrapped, wrote that the GWR intended to purchase No. 5, referring to the locomotive again as *Shannon*, and exhibit it on 'The Lawn' at Paddington station under cover. The plan did not, as posterity shows, come to fruition. The Great Western did still buy the locomotive. The possible fate of the old engine had come to the attention of Charles Collett, the GWR's chief mechanical engineer (CME) at Swindon. Collett, familiar with the locomotive during his time both as Churchward's assistant and as CME in his own right, approached the company's board of directors requesting that the engine be purchased by the company for preservation. The GWR's directors agreed with Collett's opinion and contacted the auctioneers to make an offer for the engine, which was accepted. The *Reading Mercury*'s 1 June 1946 edition reported that *Shannon* had been purchased by the GWR through tender and no doubt a deep sigh of relief was heard from all those concerned about the future of the locomotive.

No. 7 at work on the demolition train in July 1946. The locomotive was used by the contractor dismantling the tramway before it was then sold on. *John Scott-Morgan Collection*

A Swindon works portrait of No. 5 after overhaul at the GWR's workshops. The engine is in a hybrid scheme, wearing Wantage Tramway livery, but carrying the name *Shannon* which it carried whilst on the Sandy and Potton Railway but never on the tramway where it was unofficially known as 'Jane' although never carrying a nameplate. *John Scott-Morgan Collection*

The very last annual general meeting of the Wantage Tramway Company on 12 November 1946 was a short and sombre affair. The directors informed the assembled shareholders that the current cash was, as suspected, only sufficient to enable the preference shareholders to be paid the full value of their shares, which would be done once the company had been closed down. The ordinary shares, as advised in the January meeting would probably be only partly paid, the exact amount to be notified once the remaining assets of the company had been sold and the final accounts of the company had been produced. The shareholders voted to accept whatever payment was available from the money held by the tramway when the company was closed. One last act needed to be completed, the formal liquidation and closure of the company and the appointment of two liquidators to manage the final days of the company and the disposal of all its remaining assets. A formal Extraordinary General Meeting was required for the vote to take place, which was agreed to take place two weeks after the annual meeting, on 26 November.

The short Extraordinary General Meeting duly took place, with a single motion to be voted on. The motion was proposed 'that the company would be wound up voluntarily and that Percy Slone Clark of Town Mill in Wantage in the county of Berkshire, Merchant Justice of the Peace and Henry Saunders of Mount Pleasant House Wantage aforesaid Gentleman, be and they are hereby appointed Liquidators for the purpose of such winding up'. The resolution was passed and the company officially ceased trading and went into liquidation.

One last act remained, the presentation of the final accounts of the liquidated company at an Extraordinary General Meeting. The meeting was scheduled for 15 April 1947 in Wantage and was well attended. The liquidators, Clark and Saunders, presented the last accounts of the Wantage Tramway Company which were voted on by the assembled persons and passed. Clark informed the gathering that all cheques settling debts to the company had been presented to the bank, and all the tramway company's debts had likewise been settled, with all the cheques having cleared the bank. No liabilities or creditors now existed for the tramway. Payment of the full value of the preference shares had been made to the holders of the shares, and the cheques had all been paid by the bank. The holders of ordinary shares had been paid and the cheques had likewise been cleared. There were now no assets left at the bank, the balance of the company's bank accounts standing at nil. The work of the liquidators was done apart from one final sad act. One motion remained to be passed by the assembled shareholders, the proposal being 'that the books and documents in the custody of the Liquidators belonging to the Company be held by the Liquidators for a period of six months and afterwards destroyed'. The motion was passed and the meeting was called to order. The assembled people slowly and sadly left the hall, followed by Clark and Saunders, and as the latch on the door clicked shut, the Wantage Tramway Company finally ceased to exist.

The company refused to completely cease to exist though. The liquidators fortunately opted not to destroy the papers in their possession as instructed but instead opted to pass the files into the care of Berkshire County Archives in Reading, and the legacy of the tramway lived on.

Once the GWR had completed the cosmetic restoration of the engine it was sent back to Wantage Road where it was placed on a plinth in a shelter on the Down platform. The plinth on Wantage Road station wasn't quite the secure home that No. 5 needed, British Rail announcing plans to close Wantage Road station in 1964 with the locomotive needing to find a new home urgently. The correspondence files of British Rail's Public Relations and Publicity Department contain a large file of letters relating to the future of No. 5, almost always referred to as *Shannon*. Wantage Urban District Council wrote to British Rail Western Region's PR and Publicity Officer N.T. Lavenbury, on 2 February 1965. In the letter the council stated that they wanted to keep the locomotive in Wantage and asked for the terms under which the

No. 5 undergoing restoration at Swindon Works, 1946. *John Scott-Morgan Collection*

railway would be prepared to offer it to the council, although they wrote that they hoped that British Rail would be prepared to give No. 5 to the town in view of its important local significance.

The same day, a news article appeared in the *Swindon Evening Advertiser* reporting on the discussions at the previous day's council meeting which revealed the intended future of *Shannon*. The reporter wrote that the council reported that it had been told by British Rail that the locomotive could not be accommodated in the railway museum in Farringdon Road and would most likely be found a place in the intended museum at Stratford in London. A 'serious suggestion' had, as a result, been made by councillor Ivor Everson that the locomotive ought to be kept in the town. His plan was that it should be stripped of any 'dangerous' articles and then placed in the open air in an adventure playground in Wantage. Councillor Miss R.T.P. Whipple had responded by suggesting that perhaps a better solution would be for local railway enthusiasts to do something to keep the engine in the vicinity where it could be cared for rather than leaving it out in the open where it would deteriorate.

British Rail's PR department was not impressed by the plan to put the locomotive in a playground and referred the case to the Chief Publicity Officer on British Railways Board, Eric Merrill, and the Curator of Historical Relics, J.H. Scholes under whose remit the locomotive now sat. An internal memo dated 9 March 1965 from Scholes to Merrill, makes it clear that there was great concern about the council's possible plan for the locomotive. The locomotive was on British Rail's official preserved list and had been important part of the railway's national collection of historical relics since the department had been formed in 1951. Scholes stated that the locomotive would be removed when Wantage Road closed and then be stored at Stratford under museum standards. The approach by the urban district council had caused him great concern, as there was no museum which could accommodate the engine, and it seemed unlikely to the curator that the council would be able to raise funds to construct such a building nor to maintain the locomotive 'to our standards'.

The memo ended with a recommendation that the Public Relations department should write to Wantage Urban District Council offering the locomotive on indefinite loan on the condition that proper housing be provided for it in a museum where it could be placed on public display and that it was maintained to British Rail's museum standards. If, however the council was planning on placing it in a playground, Scholes advised that the approach should be declined completely and that he would be unprepared to release *Shannon* from his department's care. The letter was sent to the council by Merrill accordingly.

The council's response on 24 July stated that the adventure playground proposal had never been taken seriously by the councillors and that they were proposing to display the locomotive in a new car park which was to be built in Limborough Road. Consideration as to how the locomotive should be housed in the car park was to be given once British Rail's terms had been received. The proposal was not received with any enthusiasm within British Rail. The approach was rejected and after a period of time in store at the Atomic Energy Establishment at Harwell, where some members of the Great Western Society worked, a suggestion was made that a new home was needed for No. 5. Didcot was selected as the home for the British Transport Commission, now the national collection, engine. Didcot made a lot of sense as the new home for *Shannon* as the closest railway heritage centre to its old home at Wantage, and the locomotive remains there to this day.

No. 7, unwanted by the GWR and now redundant after it had completed the lifting of the tramway, was purchased by Newport Cordes Steel Mills Limited and moved to their Dos Works steel mill in Newport, South Wales for use working the internal railway network in the mill. The locomotive was purchased from Wantage and moved to Wales where it was overhauled and renamed *Coffeepot*. The locomotive remained in service until the 1960s when it became surplus to requirements and

Safe at last. No. 5 at Didcot Railway Centre accompanying *Bonnie Prince Charlie* in May 1969. Both locomotives are still at Didcot in 2024. *John Scott-Morgan Collection*

was offered to preservationists. Unfortunately, there was no interest from the early preservation movement despite the engine being ideal for the short lines then common on Britain's heritage railways. The lack of interest may have been due to the mechanical state of the locomotive after a long working life, with other industrial locomotives being available from the National Coal Board and directly from British Rail. The locomotive, with no prospect of preservation, was scrapped, leaving only No. 5 to carry the flag for the Wantage Tramway.

Meanwhile, the erstwhile shareholders of the company gathered for one final time on 15 April 1946 in Wantage for the Return of the Final Winding Up to be presented and approved. At the end the company had funds of £9,415.0s.5d following the sale of all its assets and the collection of the last debts owed to the company. The cost of closing down the company had comprised £46 6s 6d for liquidators' solicitors fees and a further £8 2s 6d for the cost of placing the required notices of closure in

No. 7 at work on the internal railway at Caros Droes Works, Newport in 1960. *Vale and Downland Museum, Wantage*

the *London Gazette* and the local newspapers. Expenses payable to the liquidators amounted to £13 1s 8d. The company had outstanding debts to creditors totalling £386 15s 5d which had been paid off leaving a final sum to be paid to shareholders for their shares. Preference shareholders were paid up in full, with each £5 share being paid. The full £3,000 value of preference shares was thus cleared. Ordinary shareholders did not receive the full face value of their shares, the remaining funds allowing for a payment of £3 12s 11½d to be made for each share. The final £5,960 14s 4d of company funds was paid out, and the company formally wound up.

The Wantage Tramway Company had legally ceased to exist after seventy-two years.

Chapter 15

THE LOCOMOTIVES OF THE WANTAGE TRAMWAY

NO. 1: THE GRANTHAM PATENT STEAM TRAMCAR

The Grantham Patent Steam Tramcar was the first steam powered vehicle to arrive on the Wantage Tramway and was thus given the number 1, although the first horse drawn tramcar (later converted to be hauled by steam locomotives) was also assigned No. 1.

The patent for a self-propelled tramcar using its own steam engine located on board the tram was lodged by John Grantham with the experimental design then

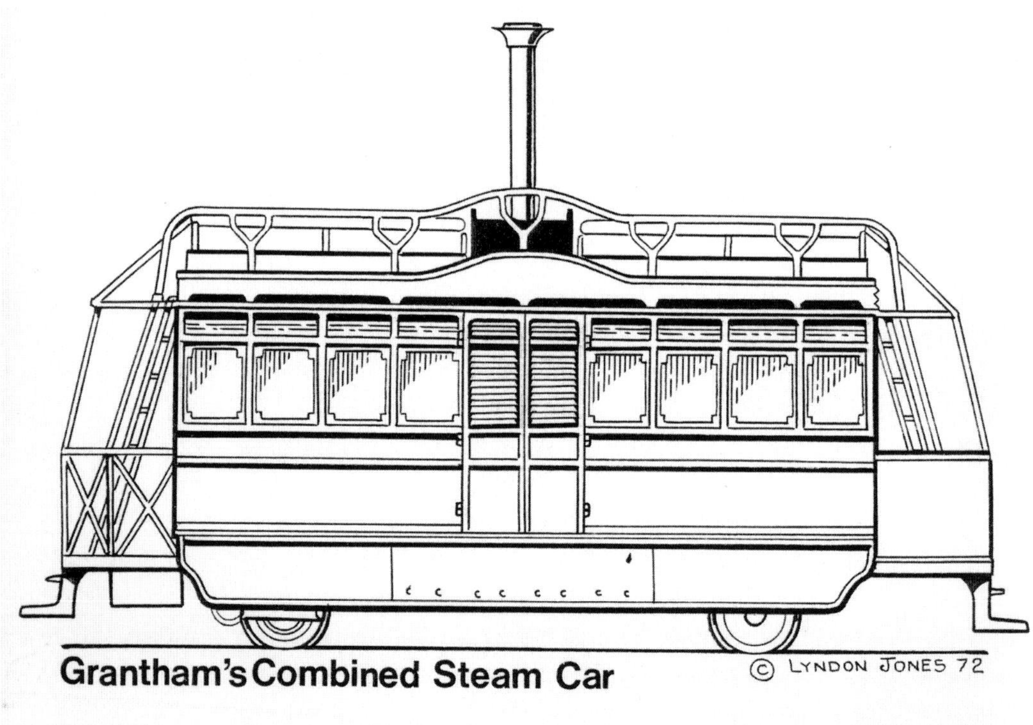

Line drawing of the Grantham Patent Steam Tram by Lyndon Jones. *Vale and Downland Museum, Wantage*

being built by the Oldbury Carriage Works in Oldbury between Smethwick and Dudley. The steam tramcar, which used an engine and boilers built by Merryweather and Sons of London was completed in 1872. Reports that the tramcar was being supported by Robert Loyd Lindsay and that his financial support was the reason for its arrival on the Wantage Tramway are not true. Loyd Lindsay does not appear to have had any involvement with Grantham's tramcar project, the archival evidence showing that the Steam Tramcar was a venture purely driven by John Grantham, who funded most of the work as he believed his patent would result in licence royalties for the construction of fleets of the trams for all Britain's tramways and those in Europe. This was not to be, No. 1 was the only Grantham Patent Steam Tramcar ever built.

The Patent Tramcar was a double-deck vehicle with seats for twenty passengers inside on the lower deck, and another twenty-four outside on the upper deck either side of the clerestory which allowed light to enter the lower decks as well as the side windows. The smoke stack protruded through the centre of the floor of the upper deck and it can only be imagined how much of the smoke and smuts emanating from the chimney were covering those unfortunate enough to have opted to travel on the upper deck. Access to the upper deck was via a vertical iron ladder at each end of the tramcar.

The tramcar was 30ft long and travelled on two fixed axles located towards the middle of the car with a wheelbase of only 10ft between the centres of the 18in diameter wheels on each side. Steam was generated by two 4ft high, 18in diameter boilers located centrally on either side of the car, both of which were enclosed allowing a gangway between them through which the conductor could pass when selling tickets, and keeping them away from the passengers on the lower deck. The power was applied to only one axle and was generated by machinery underneath the floor of the tramcar which comprised two small 4in diameter cylinders with a 10in stroke.

Trials of the Patent Steam Tramcar were undertaken by John Grantham on a short section of railway track in West Brompton in London during 1873. The car was used experimentally on the London Street Tramway along Vauxhall Bridge Road between Victoria railway station and the Vauxhall Bridge. Despite being used on London's tramways, the tramcar was merely undertaking trials and was not in revenue earning commercial service. Only when it was in use at Wantage was it carrying fare paying passengers, and thus the Wantage Tramway became the first tramway in the world to use steam haulage. The trials proved unsuccessful and the tramcar was withdrawn but the failure was due to the nature of street tramways rather than an inherent fault in the design.

The Grantham Tramcar operated well on the railway track that had been used at West Brompton. It is likely that the 'short length of railway track' referred to in contemporary reports was the District Railway's branch to Gloucester Road which at the time terminated at West Brompton. The other West Brompton station was that belonging to the West London Extension Joint Railway on the extension line from Kensington (Olympia) and Clapham Junction. It is possible that either line was used but trials on the short District Railway seem more likely.

The issue with Grantham's tram seems to have arisen only when the tram was used on the London Tramway. The grooved tram rails embedded into the roadway caused a large amount of friction on the wheels and the rigid wheelbase of the car caused difficulty in negotiating the sharp curves to be found on a road tramway. It also proved impossible to replenish the fireboxes of the two boilers with test 'passengers' on board and so no further use was seen for it in London.

The Wantage Tramway was built differently to the urban street tramways and was more akin to the railways which it had initially been trialled on. The Wantage Tramway Company purchased the Grantham Tramcar, but it was not a great success. The Grantham car had been considerably rebuilt prior to its purchase with the two boilers being replaced by one single larger boiler placed centrally but to one side to allow for a gangway. The wheels which had caused so much trouble in London were replaced by larger 24in diameter wheelsets. Contemporary accounts comment on changes made to the driving arrangements on the tramcar, stating that after the rebuild it was possible to drive the Grantham tram from each end, whether this refers to improvements to an existing arrangement or the addition of a new driving position at one end where one had not previously been is not clear.

The rebuilt car was 27ft 3in long and 6ft 6in wide. The tramcar is reported as having a height of 11ft 1in after the rebuild and a weight in operational order of 8 tons. The passenger accommodation had been increased to allow the car to carry sixty passengers, but no mention is made of any improvements to the means of accessing the upper deck, so it seems that the vertical ladder remained the only means of getting between the upper and lower decks.

The rebuild and the improvements which followed as a result of Thomas Nalder's inspection and the work undertaken by Nalder and Nalder in Challow, and Avonside in Bristol, did not make the Grantham tram a great success. Even with improvements to the boilers and draughting, the Grantham tram was not able to generate enough steam to reliably negotiate the gradients on the line and so it was replaced by a Merryweather Tram locomotive.

The suggestion that the tramcar was sold to the Portsdown and Horndean Tramway, with it reappearing in service on that line in 1903 and being found there derelict when the tramway closed in 1934 as made by R.W. Kidner in 1947 and H.A. Whitcombe in 1954 has led to debate as to the fate of the Grantham car. The Wantage directors' meeting minute book states that Grantham tram had been stripped of its boiler and machinery with the remaining car being used briefly as a locomotive hauled tramcar before being left derelict at Wantage. It was finally broken up and was not sold to another tramway.

NO. 2: ENGINE NO. 2

Locomotive No. 2 was a Merryweather Tram locomotive. The cylinders of 6in diameter and with a 9in stroke, 2ft diameter wheels and connecting rods were enclosed by steel skirts with the boiler, firebox and drivers cab all enclosed in a wooden structure with windows along each side and at each end. The locomotive had been built by Merryweather and Sons of Greenwich Road, London in 1876.

The tram locomotive was originally built for trials on the London Street Tramways, before being sent to Wantage for trials, eventually being purchased by the Wantage Tramway Company. The Merryweather Tram was used by the tramway company on its passenger services lasting until being replaced by No. 6 when it arrived.

NO. 3: ENGINE NO. 3

Locomotive No. 3 remains an enigma although the records of the Wantage Tramway in the Royal Berkshire Archives contain clues which reveal its identity. Little has been mentioned of it, although it definitely arrived on the tramway and was used. In Colonel Hutchinson's initial report written after his first visit to certify the tramway to be fit to have steam locomotives working it, he mentions that he was to postpone the final report until *one* of the Merryweather Engines had arrived at Wantage, at which point he was to return and complete his survey. It seems clear that No. 3 was a second Merryweather Tram Locomotive, and would have been identical to No. 2, unless improvements had been made in the design between the construction of the two engines. Whilst this is possible, no accounts of any differences between the two tram engines have come to light, and it would seem safe to assume that they were twins.

NO. 4: ENGINE NO. 4

Locomotive No. 4 was a closed in tram locomotive designed in 1876 by Henry Hughes and Company of the Falcon Works in Loughborough. The Hughes company would eventually become the Brush Electrical Engineering Company and would go on to be part of the Hawker Siddeley Group of engineering companies.

No. 4 was built to the Hughes patent for tramway locomotives and was one of many which were to be used on the street tramways in Glasgow, Swansea, Bristol, Guernsey and Lille in France during the 1880s and 1890s. It was a four coupled 0-4-0 saddle tank boxed out in a tram style, with all of the wheels, coupling rods and cylinders covered with steel plates. The boiler, firebox and driver's cab were also boxed in with a large wooden structure with windows along both sides, giving it the appearance of a large garden shed. The locomotive arrived on the Wantage Tramway in 1877 to work the passenger service.

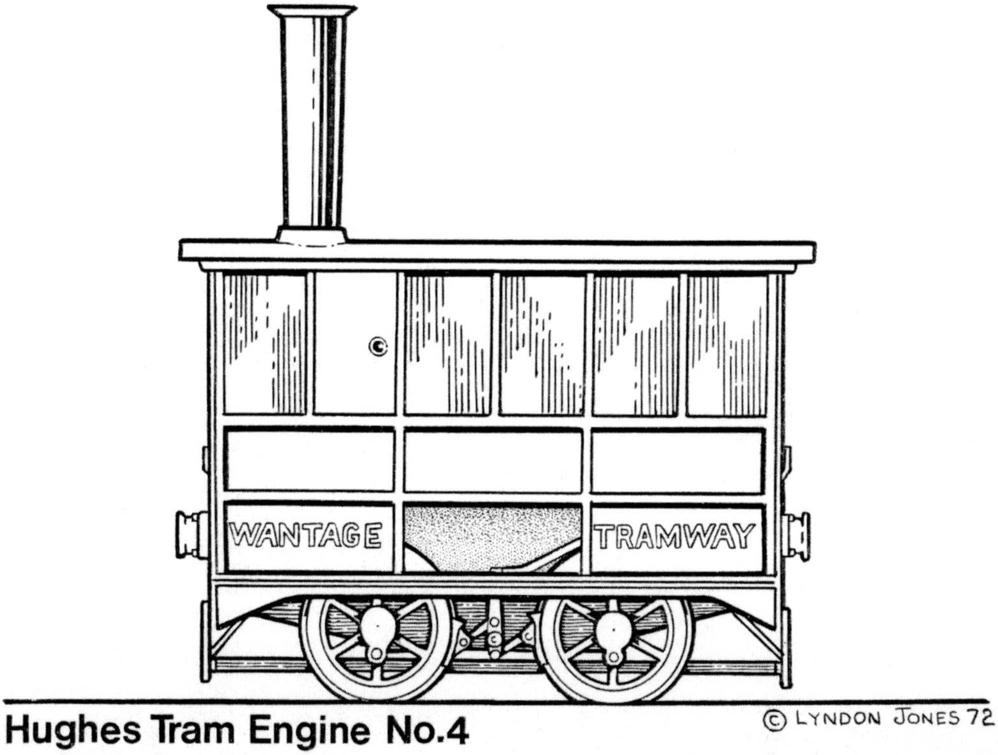

Line drawing of the Hughes Steam Tram Engine No. 4, by Lyndon Jones. *Vale and Downland Museum, Wantage*

The cylinders were of 7in diameter with a stroke of 12in and the wheels of 2ft 6in diameter. The coupling rods were removed at some point during the 1880s and the locomotive continued to run with the cylinders driving just one axle from then on until it was decommissioned by the tramway.

NO. 5: *JANE*

Locomotive No. 5 was built by George England and Company in Hatcham, New Cross, London. The company supplied many of its locomotives to contractors building a variety of railways, canals and dockyards. It also supplied locomotives to the Ffestiniog Railway, the Caledonian Railway in Scotland, the Great Western and even supplied locomotives to places as far away as Australia.

Wantage Tramways No. 5, named *Jane* by the tramway, was a four coupled 0-4-0 well tank locomotive, with the water tank located between the frames underneath the boiler. The engine had worked during a long career before being purchased by the Wantage Tramway Company.

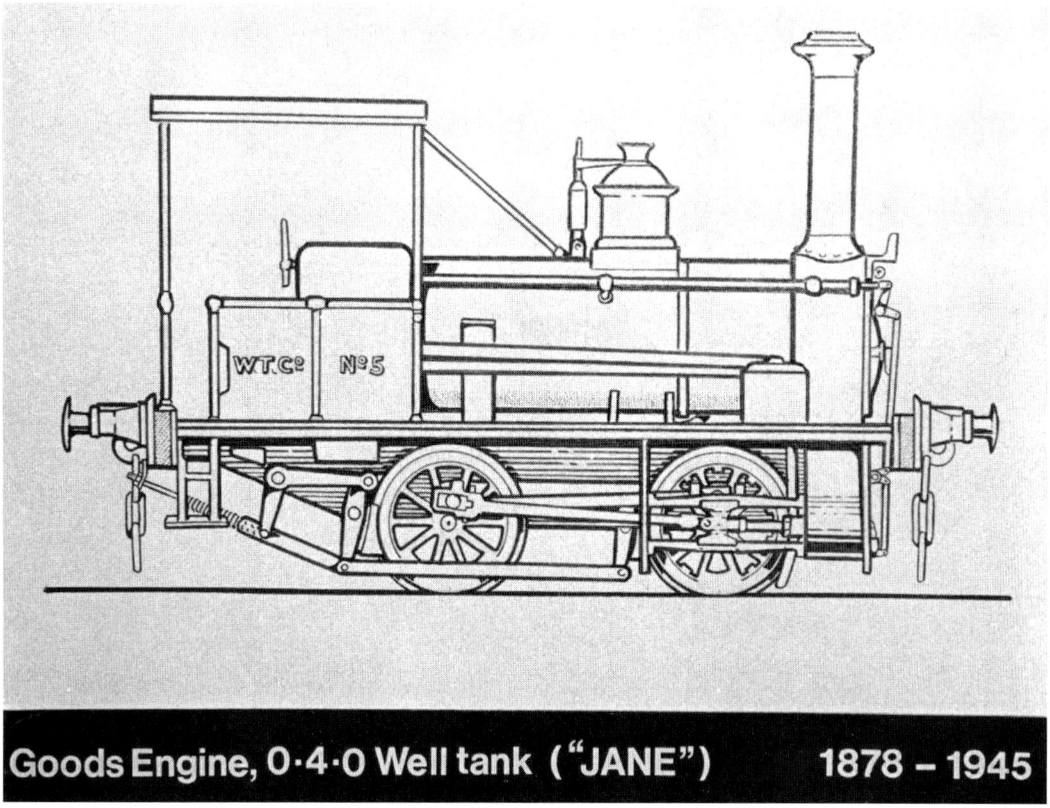

Line drawing of George England Tank Engine, number 5, 'Jane', by Lyndon Jones. *Vale and Downland Museum, Wantage*

The newly constructed locomotive, the only one ever owned by the railway, was built at a cost of £800 for Captain William Peel, the owner of the Sandy and Potton Railway in Bedfordshire in 1857 to work the line when it first opened. The railway ran mainly through Peel's family estate and was a private venture.

The locomotive was named *Shannon* by Captain Peel's mother, Lady Peel, in his absence while he was serving abroad. It was named after the Royal Navy Frigate which Peel had commanded in the far east, operating in Chinese waters. Peel, the son of Robert Peel the founder of the modern police force, never saw his railway in operation. Captain Peel was ordered to Calcutta at the start of the Indian Rebellion in 1857 and was knighted in recognition of his distinguished service in the Royal Navy shortly after his arrival. Whilst in India, Captain Peel contracted malaria and died in Cawnpore during 1858 aged only 33.

The Sandy and Potton Railway was acquired by the London and North Western Railway in 1862, who extended the route, it becoming part of the Bletchley to

Cambridge branch. Inherited by the LNWR and no longer fit for use on the longer Cambridge branch the locomotive, No. 1104, was trialled unsuccessfully on the steeply graded Cromford and High Peak Railway. The LNWR then moved the locomotive to Crewe where it was renumbered 1863 and given the name *Shannon*.

The locomotive was surplus to requirements at just the time that the Wantage Tramway Company was looking for an engine to work the line. The locomotive was sold to the tramway for just over £365 in 1878 and moved under her own power from Crewe to Oxford and on via the GWR to Wantage Road. On arrival the locomotive was given the No. 5 in Wantage Tramway stock and unofficially named *Jane*.

The locomotive was little altered over time, although it received new cylinders in 1882 to replace life expired ones, and also received a new boiler at Swindon works in 1897 to replace its original which had become beyond repair. Initially built without one, the locomotive received a cab, which it still carries, at some point in its ownership by the LNWR.

The LMS magazine of June 1937 reported that the locomotive had been rebuilt by the LNWR who added not only its cab at Crewe but also had the buffers raised to 'standard height' and its boiler water feed pumps replaced with LNWR injectors.

The locomotive continued to serve the tramway until very nearly the end. It was reported as being out of use in a semi dismantled state outside the locomotive shed at Wantage Town when *Railway Magazine* reporters visited the tramway in 1930. It was, after a campaign to preserve the engine when the tramway closed in 1946, purchased by the GWR for preservation, cosmetically restored at Swindon, and displayed at Wantage Road station. The locomotive was ultimately inherited by British Rail as part of its historical collection. The closure of Wantage Road in 1965 meant that *Jane* needed a new home. The locomotive was moved into temporary store at the Atomic Energy Authority's facility in Grove where it was looked after on British Rail's behalf.

Jane came to the attention of some members of the Great Western Railway Society who arranged for the locomotive, by then part of the British Transport Commission collection, to be moved to Didcot Railway Centre with a view to restoring the engine to steam. This was achieved, with *Jane* steaming once more in October 1969. Sadly, serious cracking was found in the firebox which would require major to work to return the locomotive to steam.

Now part of the National Collection, the locomotive can be found at Didcot Railway Centre on static display. No. 5 is one of the very few relics from the tramway, which alongside the old bus garage from Wantage Town station (also at Didcot), and the tramway offices in Mill Street, Wantage, survives today. Only a few other tiny relics remain to show that the tramway ever existed.

The locomotive had outside cylinders of 9in with a 12in stroke. The four coupled wheels were of 3ft diameter and arranged to provide a wheel base of 6ft. The boiler was of 2ft 10in diameter, with a working pressure of 120lb / sq in. The locomotive's firebox was 1ft 10 in in length, with a width of 3ft 1in to allow it to fit between the locomotive's frames. Overall, the locomotive weighed 15 tons, and had a height of 11ft 9in and a width of 7ft 8in.

NO. 6: THE MATTHEWS TRAM

Locomotive No. 6 was a closed in tram locomotive built to a patent owned by James Matthews of Bristol in 1879. The locomotive was most likely built by Fox Walker and Company also of Bristol and based at the Atlas Works. The company was purchased by Thomas Peckett in 1880 becoming Peckett and Sons.

The locomotive is believed to have been unique and so it is likely that the Matthews Tram was trialled on the Liverpool Tramways system, possibly during 1880. A contemporary journal article described a locomotive identical to the Matthews Tram being tested in Liverpool, but whether this was by the municipal tramway company or by James Matthews is unclear. The Wantage Tramway Company purchased the locomotive for use on passenger trains, and thereafter referred to it as 'The Passenger Engine'. The locomotive remained the company's main passenger locomotive from its arrival until the service ceased on 31 July 1925 when it was laid up and then put up for sale.

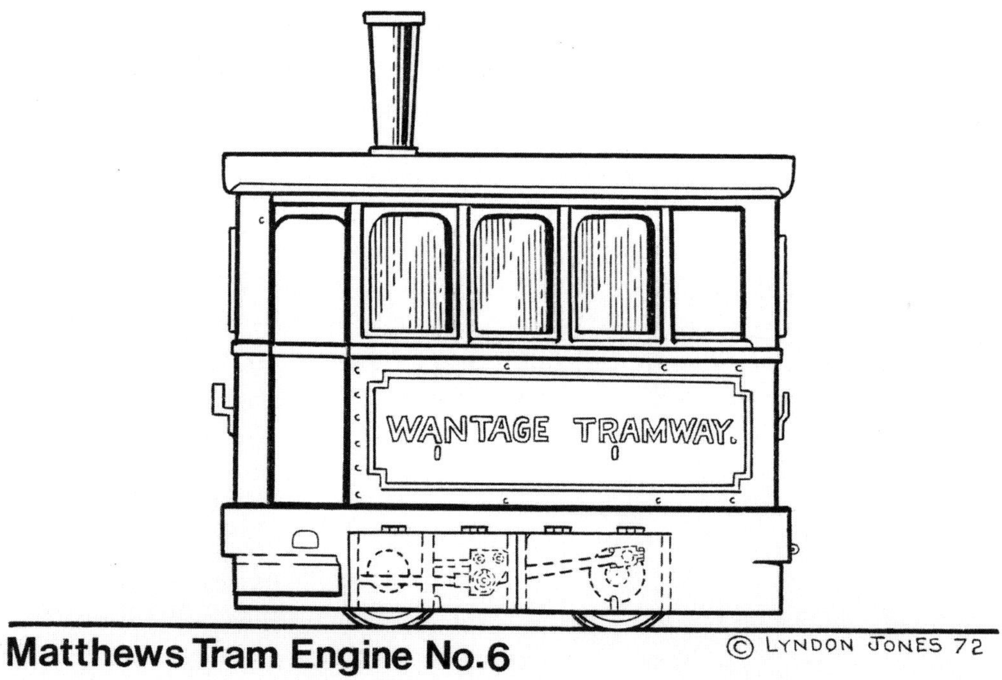

Line drawing of the Matthews Steam Tram Engine, No. 6, by Lyndon Jones. *Vale and Downland Museum, Wantage*

The Matthews Tram awaiting departure from Wantage with a well loaded passenger service. The platform is at the opposite side of the tram car, passengers supposedly not to board the cars from the near side. *John Scott-Morgan Collection*

No. 6 was an entirely closed in locomotive with the wheels, outside cylinders and connecting rods all covered with plates on the outside of the engine. The boiler with a working pressure of 120lb / sq in, firebox and cab were all located within a plated in box in the manner of many more modern steam tram locomotives. The four wheels of 2ft 4in diameter, were uncoupled, only one axle being driven by the 6in diameter cylinders which had a 10½in stroke. The locomotive was 11ft long, 7ft 3½in wide, with a height of 12ft 10in. The locomotive weighed 8½ tons in working order.

In a 1929 article presented in the *Railway Magazine* the locomotive was stated as being out of use and for sale as scrap. A later article in the *Railway Magazine* in 1936 stated that the Matthews Tram had been broken up before the author's visit after lying derelict at Wantage Town depot for some time.

NO. 7: ENGINE NO. 7, *MARY*

Locomotive No. 7 was built by Manning Wardle and Company in Leeds in 1880. The Manning Wardle Locomotive Building List shows the engine as works No. 1057 which was built for the contractor T.A. Walker for use in building the Manchester Ship Canal. When the engine was delivered to the company it was named *Brassey*, a name it kept until it was sold. Once the ship canal had been opened the locomotives

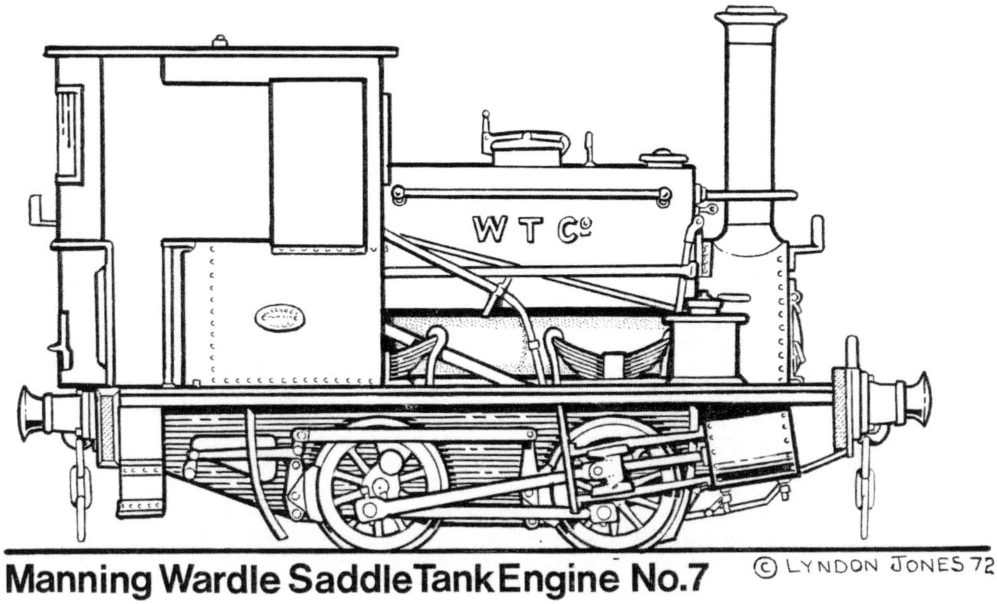

Line drawing of Manning Wardle Tank Engine No. 7, *Mary*, by Lyndon Jones. *Vale and Downland Museum, Wantage*

became surplus to requirements and were sold off by the contractor. The Manning Wardle list then shows the engine as having been sold on to the Wantage Tramway Company who purchased the locomotive directly from the contractor to replace another worn out locomotive in 1888, with it receiving the unofficial name *Mary*.

No. 7 was an 0-4-0 saddle tank with outside cylinders of 10in diameter with a 14in stroke. Its four coupled wheels were of 2ft 9in in diameter. Originally built without a cab, it received a spectacle plate during its life as a contractor's locomotive. A full cab was added to the locomotive later in its life by the Wantage Tramway Company.

Despite reports to the contrary the locomotive was not scrapped, or sold for scrap, by the Wantage Tramway Company when the tramway closed. The locomotive was purchased by Newport Cordes Steel Mills Limited and moved to their Dos Works steel mill in Newport, South Wales, where she saw out the remainder of her working life. The locomotive was offered to preservationists when it had become surplus to the needs of the steelworks in the late 1960s but it was sadly refused and the only other surviving Wantage locomotive was scrapped.

THE GREAT WESTERN ENGINE, 1329: *RAVEN*

In 1910 the tramway was urgently looking for a locomotive to operate the goods services on the line. With both Nos 5 and 7 being in urgent need of heavy overhauls another engine was desperately needed. Enquiries made to the GWR at Swindon

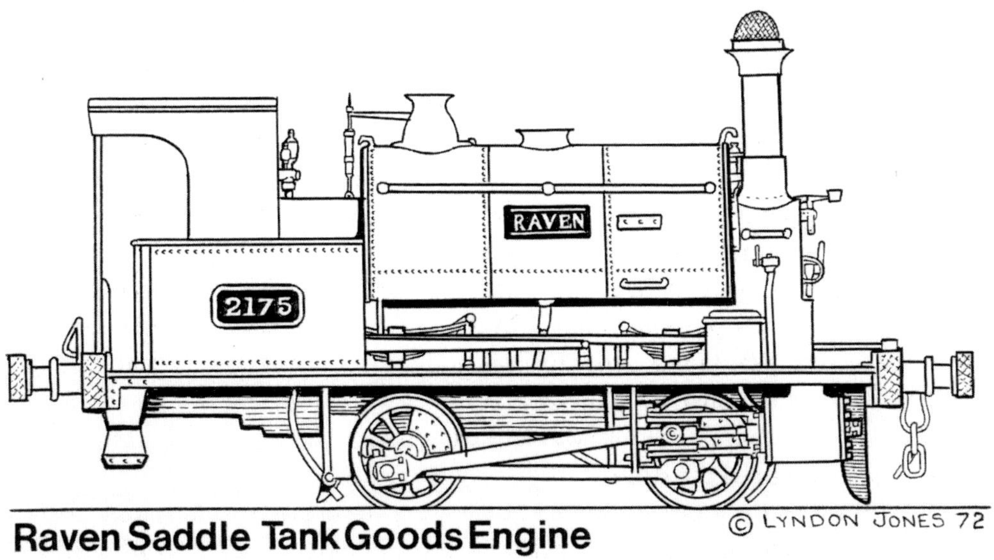

Line drawing of Tank Engine No. 2175, *Raven*, by Lyndon Jones. *Vale and Downland Museum, Wantage*

revealed that there was a locomotive there which the company was prepared to sell to the tramway.

The locomotive was No. 1329 which had been part of a batch of five identical engines built by Avonside in 1874 for the South Devon Railway. They were small 0-4-0 saddle tanks with 3ft diameter wheels, and 14in diameter cylinders with an 18in stroke. A feature of the locomotives when built was a polished brass dome on the boiler, but this was later painted over when purchased by the tramway.

They had been built to the broad gauge but were of 'convertible type' which was common around that time due to the planned abandonment of broad gauge. The locomotives had all been sent to Plymouth for use in shunting the dockyards there. The South Devon Railway was absorbed by the GWR in 1876 with the five tank engines being taken into Great Western stock, with the locomotive in question being renumbered 2175. The locomotive was regauged following the conversion of the GWR to standard gauge and became No. 1329.

Purchased by the tramway and named *Raven*, 1329 proved useful, although a little heavy for the tramway line. It remained at work until 1919 when it was badly damaged in a collision. The locomotive was so badly damaged as to be beyond economic repair and had to be scrapped at Wantage.

THE PNEUMATIC ENGINES

Two unusual locomotives ran on the Wantage Tramway in 1880. A trial was arranged by the British representatives of the Mékarski company in France who were attempting to sell their locomotives in Britain. The company had been granted a United Kingdom patent for their engines and thus they were keen to show them off to prospective buyers. The Compressed Air Engines Company Limited supplied two locomotives for trial on the Wantage Tramway, which were numbered 1 and 2, even though the same numbers were still being carried by the Grantham Patent Steam Tramcar and the original Merryweather Tram which continued to use them until they were taken out of service. Both the Grantham and Merryweather engines outlasted their pneumatic rivals.

The locomotives were powered by compressed air with the reservoirs being replenished at a compressor house. Mékarski offered two of their pneumatic locomotives to the Wantage Tramway Company on trial for free. They also arranged to build an engine house at Wantage Town and to supply all the compressor machinery free of charge. The innovative directors of the tramway company were intrigued and accepted the offer.

The Board of Trade report dated 16 September 1880 into the locomotives before they were licenced for use reveals the details of the engines. They were identical, both running on four wheels and powered by air motors which were fed from four large compressed air cylinders located on the footplate. The cylinders were replenished by a compressor powered by a stationery steam engine located in a purpose-built engine house at Wantage Town station yard. The locomotives were run using only three of the air cylinders, the fourth being kept as a reserve in case of emergencies.

The engines were only able to run in one direction which required that the locomotive was turned at each end of the journey. The BoT report states that this was achieved through the use of 'small turntables at the end of each journey' which had been placed in the yards at both tramway terminuses. The locomotives appear to have been quite small but they ran for three months from 22 September 1880. It seems that the locomotives themselves were reliable, no notes of failures being made in the directors' meeting minutes for the time and with a speed of 9mph appeared promising. The need to constantly provide a supply of compressed air to the engines stopped the trials being successful. The compressor engines supplied by Mékarski at Wantage Town station yard were excessively heavy on the use of coal. The pair of pneumatic engines were more expensive to use than the three steam engines that the tramway company had at the time. The trial was ended as a result with no further use to be made of pneumatic power.

Mékarski removed their two engines and the compressor machinery from Wantage and demolished the engine house. An interesting experiment was over.

Appendix 1

THE FOUNDERS OF THE WANTAGE TRAMWAY COMPANY

The Original Shareholders of the Wantage Tramway Company were a diverse group of people from across the local area around Wantage. The Annual List of Members report of March 1874 listed all of the people who had subscribed to the share offer to fund the building of the tramway. The company was funded through 2000 ordinary shares with a value of £5 each, raising the £10,000 capital allowed under the Act of Parliament under which the company was incorporated. A total of 944 of the available shares were taken up at the initial meeting held in Wantage to form the company. The subscribers of the scheme were as follows (all towns were in Berkshire):

Subscriber	Resident in	Occupation	Shares
Walter R.H. Barker	Wantage	Surgeon	20
John Brown	West Lockinge	Farmer	20
Richard J. Bromley	Wantage	Draper	10
Thomas Brown	Wantage	Gentleman*	20
John Beckett	Westcot	Carrier	10
Benjamin Castle	Charlton	Farmer	20
James Clark	Wantage	Corn Dealer	20
Henry Cook	Wantage	Schoolmaster	10
William Caudwell	Wantage	Gentleman*	30
Edward C. Davey and James H. Frogley	Wantage	Wine Merchants	20
Benjamin Down	Wantage	Innkeeper	20

Subscriber	Resident in	Occupation	Shares
Edwin Dolby	Wantage	Surveyor	4
Stephen Dixon	Wantage	Yeoman	12
Philip Gibbons	Wantage	Agricultural Engineer	20
Henry P. Gibbons	Wantage	Agricultural Engineer	20
William J.R. Gibbs	East Hendred	Farmer	20
William Gregory	Wantage	Builder	2
Charles Hedges, junior	Challow March	Farmer	10
Charles Hedges, senior	Challow March	Farmer	10
William Jackson	Wantage	Butcher	5
Martha Jones	Wantage	Spinster†	4
Llewellyn Jotcham	Wantage	Solicitor	20
Henry Kirkpatrick	Wantage	Baker	20
Alfred Kent	Wantage	Ironmonger	10
George and John Kent	Wantage	Builders	10
Robert J. Loyd Lindsay	Lockinge Park	Lieutenant Colonel	80
Thomas Liddiard	Wantage	Grocer	20
Joseph Lewis	Wantage	Chemist	20
Charles Law	Wantage	Boot & Shoe Salesman 6	
Robert Palmer	Wantage	Baker	10
Robert Pumfrey and Son	Wantage	Ironmongers	30
Bernard Pumfrey	Wantage	Ironmonger	30
John W. Reeves	Wantage	Farmer	20
Richard Reeves	Sparsholt	Farmer	20
Robert Sansum, senior	Wantage	Innkeeper	10
John Savory	Wantage	Innkeeper	8
Edward Simmons	Wantage	Fruiterer	4
Percy Smith	Letcombe Basset	Farmer	20
Lawrence Stratton	Wantage	Grocer	20
George Stevenson	Wantage	Engineer	20

Subscriber	Resident in	Occupation	Shares
Laurence W. Surmon	Wantage	Corn Dealer	20
Thomas Tosland	Wantage	Draper	20
Thomas Walker	Wantage	Farmer	20
Herbert H. Woodward	Wantage	Clerk in Holy Orders‡	2
John Wheeler	Wantage	Builder	2

* The term 'Gentleman' referred to someone of independent means who was not employed or running a business, but had their own private income from investments, land or family wealth.

† The term 'Spinster' was used by the Victorians to refer to the 'occupation' of an unmarried woman who was not working but had an independent income of their own much as a 'Gentleman' had.

‡ The term 'Clerk in Holy Orders' referred to Clergymen such as Ministers, Priests and Vicars during the nineteenth century in official documents.

Appendix 2

FINANCING THE CONSTRUCTION OF THE WANTAGE TRAMWAY

The tramway proved a popular project with the majority of the shares on offer being subscribed to by local people. The company made calls for three part payments for each £5 share during 1874 against those people who had subscribed in order to raise funds for the construction of the line as below:

Subscriber	Resident in	Occupation	Shares	Paid (£)
Walter R.H. Barker	Wantage	Surgeon	20	100
John Brown	West Lockinge	Farmer	20	100
Richard J. Bromley	Wantage	Draper	10	50
Thomas Brown	Wantage	Gentleman	20	100
John Beckett	Westcot	Carrier	10	50
Benjamin Castle	Charlton	Farmer	20	100
James Clark	Wantage	Corn Dealer	20	100
Henry Cook	Wantage	Schoolmaster	10	50
William Caudwell	Wantage	Gentleman	30	150
Edward C. Davey and James H. Frogley	Wantage	Wine Merchants	20	100
Benjamin Down	Wantage	Innkeeper	20	100
Edwin Dolby	Wantage	Surveyor	4	20
Stephen Dixon	Wantage	Yeoman	12	12
Philip Gibbons	Wantage	Agricultural Engineer	20	100
Henry P. Gibbons	Wantage	Agricultural Engineer	20	100
William J.R. Gibbs	East Hendred	Farmer	20	100

Subscriber	Resident in	Occupation	Shares	Paid (£)
William Gregory	Wantage	Builder	2	10
Charles Hedges, junior	Challow March	Farmer	10	50
Charles Hedges, senior	Challow March	Farmer	20	100
William Jackson	Wantage	Butcher	5	25
Martha Jones	Wantage	Spinster	4	20
Llewellyn Jotcham	Wantage	Solicitor	20	100
Henry Kirkpatrick	Wantage	Baker	20	100
Alfred Kent	Wantage	Ironmonger	10	50
George and John Kent	Wantage	Builders	10	50
Charles Law	Wantage	Boot and Shoe Salesman	6	30
Mary A. Lender†			20	100
Joseph Lewis	Wantage	Chemist	20	100
Thomas Liddiard	Wantage	Grocer	20	100
Charles H. Lindsay	Lockinge Park	Colonel	200	1000
Charles H. Lindsay	Lockinge Park	Colonel	100	500
Robert J. Loyd Lindsay	Lockinge Park	Colonel*	80	400
Robert J. Loyd Lindsay$	Lockinge Park	Colonel	400	2000
John McKay†	Wantage	Tea Dealer	25	125
John S. Meadows			20	100
Edward Ormond†	Wantage	Solicitor	40	200
The Right Honourable The Lord Overstone†‡	Overstone, Northants	MP	80	400
Robert Palmer	Wantage	Baker	10	50
Robert Pumfrey and Son	Wantage	Ironmongers	30	150
Bernard Pumfrey	Wantage	Ironmonger	30	150
John K. Reeves†	Wantage	Farmer	20	100

Subscriber	Resident in	Occupation	Shares	Paid (£)
John W. Reeves§	Wantage	Farmer	20	0
Richard Reeves	Sparsholt	Farmer	20	100
John Savory	Wantage	Innkeeper	8	40
Robert Sansum, senior§	Wantage	Innkeeper	10	0
Edward Simmons	Wantage	Fruiterer	4	20
Percy Smith	Letcombe Basset	Farmer	20	100
George Stevenson	Wantage	Engineer	20	100
Laurence Stratton†	Wantage	Grocer	20	100
Laurence W. Surmon	Wantage	Corn Dealer	20	100
Thomas Tosland	Wantage	Draper	20	100
Thomas Walker	Wantage	Farmer	20	100
Herbert H. Woodward	Wantage	Clerk in Holy Orders	2	10
John Wheeler	Wantage	Builder	2	10

† Denotes shareholders who had subscribed or purchased shares outright following the initial meeting to form the company.

$ The Loyd Lindsays subscribed to a total of 700 shares to provide sufficient capital for the construction of the line during 1874 after the formation of the company.

* Robert Loyd Lindsay had been promoted to full Colonel shortly after the formation of the company.

‡ Lord Overstone was the father of Harriet, the wife of Robert Loyd Lindsay and one of the founders of Jones, Loyd & Co. Bank which later became the National Westminster Bank. Overstone was an MP and a member of the British Relief Association which raised over £500,000 to support the famine victims in Ireland during the 1840s.

§ Robert Sansum had not paid the calls for his share subscription by the end of 1874 but did pay the full amount early in 1875. John Reeves had requested to give up his shares which were reallocated to other shareholders during 1874.

BIBLIOGRAPHY

Belcher, L, *Memoirs: The Wantage Tramway Company Limited* (Self Published, 1950).

Bett, Wingate H. and John C. Gillham, *Great British Tramway Networks* (Light Railway Transport League, London, 1944).

Binding, Walter, *Memories of a Railwayman* (Self Published, 1960).

Christiansen, Rex, *A Regional History of the Railways of Great Britain: Volume 13 Thames & Severn* (David & Charles, Newton Abbot, 1981).

De Courtais, Nicholas, *The Wantage Tramway* (Wild Swan, Bath, 2017).

J.G. Harrod & Co.'s Royal County Directory of Bedfordshire, Buckinghamshire, Berkshire, Oxfordshire, Huntingdonshire, and Northamptonshire, 1876 (Harrod and Company, Norwich, 1876).

Karau, Paul and Chris Turner, *The Wallingford Branch* (Wild Swan, Upper Bucklebury, 1982).

Kelly's Post Office Directory of Berkshire, Northamptonshire, and Oxfordshire, 1854 (Kelly and Co., London, 1854).

Kidner, R.W., *A Short History of Mechanical Traction & Travel, Part 6: Multiple Unit Trains, Rail Motors and Tramcars 1829-1947* (Oakwood, Godstone, 1947).

Mathias, Peter, *The First Industrial Nation: The Economic History of Britain 1700-1914* (Routledge, Abingdon, 1993).

Meacher John, 'The Oil Depot at Wantage', *British Railway Journal, 68*, pp.106-109.

Meacher John, 'The Oil Depot at Wantage: A Sequel', *British Railway Journal, 70*, pp.106-109.

Meacher John, 'The Wantage Tramway: Oil and Gas', *British Railway Journal, 73*, pp.345-346.

Mitchell, Vic and Kieth Smith, *Branch Line to Wantage: The Wantage Tramway. A Tramway Classic* (Middleton, Midhurst, 2003).

National Archives Kew, Various papers in AN 111, BT 31, MT 6, MT 56, RAIL 252, RAIL 254, RAIL 1005 and ZSPC 11.

National Library of Scotland, Ordnance Survey Six-Inch Series (1883), Berkshire Sheets XII, and XIV

National Library of Scotland, Ordnance Survey Six-Inch Series (1913), Berkshire Sheet XIV.SW, XIV.NW, and XIV.NE.

National Library of Scotland, Ordnance Survey (1944), Berkshire Sheet XIV.SW, 6" to 1 Mile.

Pearce-Higgins, S.H.. *The Wantage Tramway* (Abbey Press, Abingdon, 1958).

Savage, Christopher I., *An Economic History of Transport* (Hutchinson, London, 1966).

Simmons, Jack, *The Railway in England and Wales, 1830-1914* (Faber and Faber, London, 2008).

Simmons, Jack, *The Railway in Town and Country, 1830-1914* (Faber and Faber, London, 2008).

Simmons, Jack, *The Victorian Railway* (Thames & Hudson, London, 2009).

The Tramways Act, 1870, Available at: https://www.legislation.gov.uk/ukpga/Vict/33-34/78/contents/enacted

The Royal Berkshire Archives Reading, Various papers in D/ECO/B6, D/EX1052, D/EX2547, D/WT1, D/WT2, D/WT5, D/WT7, D/WT8, D/WT9,D/WT10, D/WT11, D/WT13, D/WT14, D/WT15, D/WT16, D/WT17, D/WT18, D/WT19, D/WT20, D/WT22, D/WT24, D/WT25, D/WT26, D/WT27, D/WT29, D/WT30, D/WT30, D/WT31, D/WT32, D/WT33, D/WT34, D/WT35, D/WT36, D/WT38, D/WT39, D/WT40, D/WT43, D/WT44, D/WT45, D/WT46, D/WT47, D/WT49, D/WTS1, D/WTS2, D/WTS3, D/WTS4, D/WTS5, D/WTS6, D/WTS7, D/WTS8, D/WTS9, D/WTS10, D/WTS12, D/WTS13, Q/RUM123, Q/RUM317, Q/RUO7.

Vale and Downland Museum Wantage, Various papers and images from the Museum collection.

Whitcombe, H.A., *History of the Steam Tram: Locomotion Papers Number Nine* (Oakwood, Godstone, 1954).

Wilkinson, Reg, *The Wantage Tramway* (Oakwood, Oxford, 1995).

Wilkinson, Reg, *Wantage Tramway* (Reg Wilkinson, Carlisle, 1974).

INDEX